Numerical Solution of
Antennas in Layered Media

ELECTRONIC & ELECTRICAL ENGINEERING RESEARCH STUDIES

ANTENNAS SERIES

Series Editor: **Professor J. R. James,** *The Royal Military College of Science (Cranfield), Shrivenham, Wiltshire, England*

* Out of print.

Numerical Solution of Antennas in Layered Media

Volkert W. Hansen

Ruhr-University Bochum, FRG

RESEARCH STUDIES PRESS LTD.
Taunton, Somerset, England

JOHN WILEY & SONS INC.
New York · Chichester · Toronto · Brisbane · Singapore

RESEARCH STUDIES PRESS LTD.
24 Belvedere Road, Taunton, Somerset, England TA1 1HD

Marketing and Distribution:
Australia, New Zealand, South-east Asia:
Jacaranda-Wiley Ltd., Jacaranda Press
JOHN WILEY & SONS INC.
GPO Box 859, Brisbane, Queensland 4001, Australia

Canada:
JOHN WILEY & SONS CANADA LIMITED
22 Worcester Road, Rexdale, Ontario, Canada

Europe, Africa:
JOHN WILEY & SONS LIMITED
Baffins Lane, Chichester, West Sussex, England

North and South America and the rest of the world:
JOHN WILEY & SONS INC.
605 Third Avenue, New York, NY 10158, USA

Library of Congress Cataloging-in-Publication Data

Hansen, Volkert W., 1944–
 Numerical solution of antennas in layered media / Vokert W.
Hansen.
  p.   cm. — (Electronic & electrical engineering research
studies. Antennas series: 8)
  Bibliography: p.
  Includes index.
  ISBN 0 471 92411 3 (Wiley)
  1. Antennas (Electronics) 2. Numerical analysis. I. Title.
II. Series
 TK7871.6.H353   1989   89–8516
 621.382′4—dc20   CIP

British Library Cataloguing in Publication Data
Hansen, Volkert W. *1944–*
 Numerical solution of antennas in layered
media.
 1. Antennas
 I. Title
 621.38′028′3

 ISBN 0 86380 087 4

ISBN 0 86380 087 4 (Research Studies Press Ltd.)
ISBN 0 471 92411 3 (John Wiley & Sons Inc.)

Printed in Great Britain by Galliard (Printers) Ltd., Great Yarmouth

v

EDITORIAL PREFACE

The advent of computers has indeed enabled antenna designers to optimize antennas of practical interest that hitherto have been too difficult to analyse due to their geometries. In many practical situations the effect on the antennas of a reflecting surface such as the earth, a weathershield, a vehicle etc has to be allowed for and this is often performed using ray optics or the geometric theory of diffraction. There is an increasing number of applications however where the antenna is embedded in a media to such an extent that the antenna characteristics are overridden by the surrounding environment; the electromagnetic heating of body tissue in cancer therapy and the remote sensing of rock strata are typical examples but some modern antennas such as printed dipoles in multilayered substrates fall also into this category. In many cases the situation can be modelled as parallel layers of homogeneous material and this has created a revival of interest in the classical interface theory of Sommerfeld and other established analytical methods. Unfortunately closed form solutions are not forthcoming and a reliance must be placed on numerical solutions. Furthermore the complexity of the wave trapping action in the layers is such that only elemental radiators like wire or patch dipoles can generally be accommodated in the modelling. Antennas in media thus present analytical complexities in the extreme and despite the well established nature of the classical mathematical formulations their numerical implementation raises many issues and difficulties. It is therefore a pleasure to introduce this timely monograph by Dr. Hansen who has specialised for some years in the numerical realisation of solutions to practical layered problems. The book gives a clear account of how to approach such calculations and the applications discussed are fascinating and are of major interest in themselves, particularly the study of transient responses in layered media. The book will be appreciated by designers concerned with practical electromagnetic layered problems and provide further encouragement to students and researchers in electromagnetics seeking advances in the solution of challenging practical engineering problems.

J. R. JAMES

PREFACE

Since the first few papers, which appeared in the early 50's to the 70's, printed antennas have covered a wide field of applications in microwave antenna techniques and many investigations have been published. However, the theoretical approaches applied so far, for example the aperture field or the cavity model method, are basically approximate methods, which may give efficient rules for the specification of the parameters in many technical problems. They may lead, however, to questionable results if, for example, the substrate of an antenna with improved bandwidth is not very thin or if the antenna is painted. This disadvantage can be removed with the help of methods which are based on the exact Green's function of the structure represented in the form of Sommerfeld integrals, leading to an exact solution of the boundary value problem. As the Green's function is the kernel of the integral equation for the antenna currents, the main problem which now arises is to find integration procedures for a sufficient fast evaluation of the Sommerfeld integrals. A careful look through the literature shows that the topic "Sommerfeld integrals" is often investigated in connection with geophysical prospecting and with the investigation of communication links above earth. From these papers we learn that the main problems with Sommerfeld integrals are similar whether a two– layered or a multilayered problem is discussed. Furthermore, the difficulties encountered in solving the integral equation for the currents turn out to be independent of the number of layers. Therefore it suggests the first step should be to deal with antennas in multilayered media and, as the second step, to treat the printed antenna we are primarily interested in as a special case of the first one. Because of this it is possible to work on very different technical problems; for example, on covered printed antennas, focusing antennas for hyperthermia, heating of plane layers in industrial production, geophysical prospecting, underwater communication etc.

All results presented in this book have been evaluated at the Institut für Hoch– und Höchstfrequenztechnik, Ruhr– Universität Bochum, most of them for the preparation of a post doctoral thesis ("Habilitationsschrift") . Many grateful thanks are due to em. Prof. Dr. H. Severin, former head of the institute, for supporting this work, for his advice and valuable comments. Several students who prepared their M.Sc. thesis ("Diplomarbeit") in the field of microwave antenna techniques at the institute also contributed to the material presented here. With special gratitude, I would like to acknowledge Dipl.- Ing. R. Körner and Dipl.- Ing. A. Janhsen, who carried out a lot of programming and computations, and Dr. W. Huhn, who supported the preparation of the English version.

Finally I would like to thank Professor J. R. James for encouraging me to do this work.

Bochum, December 1988

V. Hansen

CONTENTS

CHAPTER 1
Introduction and summary

The study of the radiation properties of antennas in plane stratified media has often proved to be very useful for working out approximate solutions for corresponding problems arising in the field of electromagnetic waves. Certainly, the most famous example is Sommerfeld's exact solution for the field of a point source located in the plane interface between two halfspaces of different permittivity and permeability, which can be used as an approximate solution for the propagation of electromagnetic waves along the surface of the earth. With the help of this solution a satisfactory discussion of Marconi's experiments became possible for the first time. Furthermore, it could be shown that the conductivity of the earth has to be finite in order to explain the directivity of the antenna used by Marconi.

Marconi had already observed that the range of his transmitter was much greater at night than during the day. It became possible to explain this effect assuming a reflecting layer in the upper part of the atmosphere, the properties of which change during the period of one day. From about the mid-twenties onwards ionosondes made it possible to determine reliable values for the electron-density profile and thus for the permittivity profile of the upper atmosphere. Since then, one could base the investigations of electromagnetic wave-propagation on the stratified model of the atmosphere.

From about the mid-thirties it became necessary to give up the picture of the earth being a homogeneous dielectric - which had been Sommerfeld's approach. This was mainly for two reasons:

- Firstly, precise information was needed about the influence of low-lying strata on the antenna field, e.g. for the planning of large broadcasting stations.

- Secondly, geophysicists were trying to explore the structure of the subsoil by measuring electromagnetic fields on the surface of the earth. This method, in addition to the conventional seismic methods, appeared to be promising as the electrical properties of geological material can differ considerably.

Apart from the d.c. method, above all the induction method was developed; in this method the influence of conducting strata on the magnetic field of an antenna is determined. The frequencies used - typically up to about 5 kHz - allow deep penetration into the surface of the earth; however, the consequence is poor resolution. Today, frequencies up to the microwave range are used for geological investigations.

The model of a planar antenna above a stratified structure is also very useful for developing suitable radiation elements for hyperthermy, where the problem of heating tissue beneath the skin has arisen. If one wants to determine the field of an antenna which lies above the skin surface the tissue can roughly be described as plane layers of skin-, fat-, and muscle tissue. Thus, the influence which the parameters of the layers and the current distribution have on the field distribution in the tissue can be discussed.

Printed antennas can be regarded as a special case of planar antennas in stratified media. Due to their great advantages - small weight, good aerodynamic properties and low costs in production - they have been widely used in the field of antenna technology during the last ten years. In order to protect them against damage, printed antennas often have to be covered by some material which, however, can be responsible for high losses, as is the case during the reentrance of a satellite into the atmosphere.

A further example for the use of the model considered here is the analysis of underground communication links and of communications links under water.

In order to determine the field of an antenna in a stratified medium one has to solve Maxwell's equations, taking into account the boundary conditions on the antenna and in the interfaces between the layers. The literature shows that the discussion of the problem posed by Sommerfeld - electric or magnetic elementary dipole above the homogeneous earth - already demands extensive theoretical effort and can be regarded as completed only since the mid-sixties. Using Sommerfeld's approach for two or more layers makes the analysis considerably more complicated. In this case the calculation can usually only be carried out by the use of high speed computers. In the field of high frequency and microwave technology one has to account for the fact that the length scales of antennas and wavelengths are of the same order. The conventional assumption of a point source has to be discarded in this case. In the present book therefore we will describe a method for the analysis of plane antennas of finite length in stratified media. It is based on the integral equation for the antenna current which will be introduced in section 2.2. This integral equation is formally equivalent to the one for antennas in free space, if Green's function for the stratified medium is used instead of the corresponding function for free space. The determination of Green's function, which will be discussed in section 2.3 in some detail, proves to be the most painstaking part of the method. As a first application we will investigate the radiation properties of printed antennas with a dielectric covering in chapter 3. In this case, the amount of work involved in discussing the equations for Green's function for the setup (three layers of different permittivity above a reflector) is still reasonable. This makes it possible to investigate the excitation of surface waves and - based on this - the radiative coupling between the elements of arrays of antennas.

In chapter 4, considerations concerning the design of focusing antennas above a biological tissue are discussed and some concrete examples are worked out.

Then, in chapter 5 the radiation pattern and the mutual coupling of antennas situated on the horizontally stratified earth are determined as a contribution to the application of electromagnetic methods for surveying subterranean inhomogeneities.

The extension of the method to transients will be discussed in chapter 6. It will be shown that a solution can be given in principle but it would use up too much computer time. For the special case of a straight dipole of finite length above a stratified medium, however, suitable approximations can be made and thus the farfield can be determined for the case of an excitation by a band-limited voltage pulse.

Each of the chapters 3 to 6 contains an introduction where the historical development is outlined and where we will define the problem more precisely. It proved to be too extensive to give a complete list of the literature for each of the four subjects. Therefore, we will mainly quote those papers which demonstrate the development of the analytic methods.

CHAPTER 2

Theory

2.1 Introductory remarks

The present boundary value problem can be split into two parts: first, the required solution of Maxwell's equations must fulfil the boundary conditions in the observation points on the surface of the antenna; second, it must obey the wave equation in the layers and boundary conditions in the interfaces. The strategy which is presented here is guided by this division: As a first step, an integral equation for the antenna current is derived. Thereby the procedure is similar to that of the thin-wire-theory. This means that, first of all, Green's function of a space is defined formally, with the help of which an equation for the electric field in an observation point can then be given. The desired integral equation is obtained by putting the observation points onto the antenna surface and requiring that the boundary conditions be fulfilled there. The second step involves the determination of Green's function of a stratified structure. As has been mentioned, this step is very cumbersome. The presentation which has been chosen here actually follows the one which has already been proposed by Sommerfeld. Sommerfeld has expressed the field of a point source in terms of improper integrals over solutions of the wave equation in cylindrical coordinates. Thus, he has arrived at a presentation which allows taking into account the boundary condition in the interface between two half-spaces. It is still applicable, if, instead of a single interface, one is dealing with an arbitrary amount. It has proven practical, however, not to work with Hertz' potentials, as Sommerfeld did, but to split the complete field into a TM and a TE part.

The solution of Sommerfeld's integrals has turned out to be very complicated. Since their formulation at the beginning of this century, a great number of studies have been dedicated toward this end. Some took up Sommerfeld's procedure and modified it, others introduced new procedures of solution. For the solution of the integral equations and the determination of input impedance and radiation field of antennas, it must be possible to calculate Green's function in the entire space. This is only possible if use is made of different procedures.

Especially adequate are the saddle-point method for the calculation of the farfield, the use of image sources for observation points in the nearfield and numerical integration schemes. The application of all these procedures requires a detailed knowledge of the analytical properties of the integrands. These are non-unique complex functions. Before the integration can be carried out, uniqueness has to be obtained with the help of physical and/or mathematical considerations. For the saddle-point method this is achieved via a suitable transformation and a branch cut. The integral which has been obtained in this manner can be solved analytically for large separations of source and observation points after a shift of the path of integration. Thereby one only takes into account the essential contributions, i.e. of the so-called saddle point, of the poles which have been passed during the shift of the path of integration and of the branch cut. The terms obtained in this way can be interpreted in physical terms: the contribution of the saddle point corresponds to that part which can be understood in terms of geometrical optics. The contributions of the poles describe the fields of surface and leaky waves, and that of the branch point represents lateral waves. For the examples which are discussed in this book, the latter either do not appear at all or are negligibly small.

A numerical integration of Sommerfeld's integrals is appropriate in the case where the distance between source and observation point is of the order of the wavelength. The execution of the integration with "brute force" is, indeed, usually possible, but often very time consuming. A careful examination of the integrands shows that single terms can be split off in such a way that one can find analytical solutions for the poorly converging parts, and only the remaining well–behaved integrals have to be solved numerically. Note, however, that singularities may be on the path of integration. The contribution of these singularities must be determined with the help of analytical methods.

If source and observation points are close together, a numerical evaluation of the integrals usually becomes impossible due to the field singularity in the source point. It can be shown that in this case the result is mainly determined by the behaviour of the integrand for large values of the integration variable. Keeping this in mind one can derive suitable approximate equations, where the contributions of the single terms can be interpreted with the help of image sources.

2.2 Integral equation for the antenna currents

2.2.1 The model

A plane medium consisting of $(\hat{m} + n + 1)$ layers (Fig. 2.1) is considered. Each layer consists of homogeneous isotropic material with linear properties, described by the dielectric constant ε_{ri} and the relative permeability μ_{ri}. The 0-layer contains the origin of the coordinate system. The layer of index $\hat{m}$ extends to infinity with $z > 0$, the n-layer to infinity with $z < 0$. The interface between the layers i and $(i+1)$ is formed with the plane $z = d_i$ $(z > 0)$ or $z = -d_i(z < 0)$ respectively. The plane $z = 0$ contains the antenna, consisting of an electric conductor. The thickness of this antenna will be neglected. This is why the current on the antenna can be viewed as the surface current

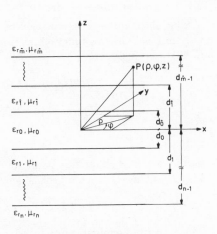

Fig. 2.1 Coordinate system in a stratified medium

$$\vec{J}_F = \vec{u}_x J_{Fx} + \vec{u}_y J_{Fy} \quad . \tag{2.1}$$

For the electric field on the surface of the conductor it is assumed that the field component tangential to the surface $\vec{E}_{tan}$ is so small that it can be neglected and thus the boundary condition

$$\vec{E}_{tan} = 0 \tag{2.2}$$

is valid.

$$k_i = \omega\sqrt{\varepsilon_0\varepsilon_{ri}\mu_0\mu_{ri}} \tag{2.3}$$

denotes the wave number of a uniform plane wave of frequency ω in the i-layer. In order to avoid the wave number for the 0-layer

$$k_0 = \omega\sqrt{\varepsilon_0\varepsilon_{r0}\mu_0\mu_{r0}} \tag{2.4}$$

being mixed up with the wave number of the free space, the latter is used without a subscript:

$$k = \omega\sqrt{\varepsilon_0\mu_0} \quad . \tag{2.5}$$

Analogously the wavelength of a uniform plane wave in free space is called λ, that of the i-layer, with the material parameters ε_{ri} and μ_{ri}, λ_i. When dealing with harmonic time variation, the time dependence is described with the factor $e^{j\omega t}$.

2.2.2 Development of the integral equation

The mathematical formulation of the problem starts from the definition of the tensorial Green's function, as it was introduced in the theory of diffraction in [2.1] and [2.2]. Let

$$\vec{I}(\vec{r}')dl = \vec{u}_x I_x dl_x + \vec{u}_y I_y dl_y + \vec{u}_z I_z dl_z \tag{2.6}$$

be a current element located at $\vec{r}'$ for frequency ω and let

$$d\vec{E}(\vec{r}) \tag{2.7}$$

be the electric field of this point source at an observation point $\vec{r}$ according to Fig. 2.1. Then Green's tensor is defined by the equation

$$d\vec{E}(\vec{r}) = \overset{\leftrightarrow}{G}(\vec{r},\vec{r}') \cdot \vec{I}(\vec{r}')dl \tag{2.8}$$

which, written explicitly in Cartesian coordinates, reads

$$dE_x = G_{xx} I_x dl_x + G_{xy} I_y dl_y + G_{xz} I_z dl_z \tag{2.9}$$

$$dE_y = G_{yx} I_x dl_x + G_{yy} I_y dl_y + G_{yz} I_z dl_z \tag{2.10}$$

$$dE_z = G_{zx} I_x dl_x + G_{zy} I_y dl_y + G_{zz} I_z dl_z \quad . \tag{2.11}$$

Thus, for the calculation of $\overset{\leftrightarrow}{G}(\vec{r},\vec{r}')$ a solution of Maxwell's equations

$$\text{rot}\vec{H} = j\omega\vec{D} + \vec{J} \tag{2.12}$$

and

$$\text{rot}\vec{H} = -j\omega\vec{B} \tag{2.13}$$

must be found for a point source. This solution must fulfil the boundary conditions at the interfaces, and the radiation condition.

The electric field for an electric current density is the result of the summation of the contribution of all point sources and is thus obtained from the integral

$$\vec{E}(\vec{r}) = \int_V \overset{\leftrightarrow}{G}(\vec{r},\vec{r}') \cdot \vec{J}(\vec{r}')dV' \quad . \tag{2.14}$$

By using the primed term dV' we want to indicate that the integration has to be performed over the sources. Because of the fact that in the model considered here all currents are running parallel to the interfaces and that thus

$$J_z = 0, \tag{2.15}$$

the components of the third column of the Green's tensor do not have to be determined. Furthermore, it results from the symmetry of the structure, according to Fig. 2.1, that the second column of the Green's tensor is obtained from the first one through a rotation of the coordinate system. Thus, the first and most complex step (as will be shown in section 2.3) for the solution of the problem is the calculation of the field of a horizontal point source in a stratified medium. Pursuing this in the following - according to the general terminology - the expressions "field of a point source" and "Green's function" will be used with the same meaning. For the formulation of the integral equation further simplifications of the model will be performed. The current is assumed as running in conducting strips with the width w, with

$$w \ll \lambda \qquad (2.16)$$

so that only the component of the current in the direction of the conducting strip has to be taken into account (Fig. 2.2). Larger conducting surfaces can be approximated with the help of a grid net of printed strips [2.3]. If one calls the surfaces of all printed circuits S and an infinitesimal element of it dS then eqn. (2.14) combined with eqn. (2.1) yields

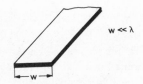

Fig. 2.2 Printed strip of width w

$$\vec{E}(\vec{r}) = \int_S \overleftrightarrow{G}(\vec{r}, \vec{r}\,') \cdot \vec{J}_F(\vec{r}\,')dS' \quad . \qquad (2.17)$$

Now the integral has to be determined as a surface integral over the printed circuits.

The printed circuits are fed from one or more sources. First of all each voltage source is modelled by an impressed electric field (Fig. 2.3). Now an integral equation can be constructed by putting the field point $\vec{r}$ on the printed circuit and by demanding that here the tangential electric field vanishes. In the feed point the field must be equal to the impressed electric field. Because of

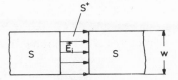

Fig. 2.3 Feed point characterised by an impressed electric fieldstrength $\vec{E}_i$

the condition (2.16) only the electric field component in the directions of the conductive strips has to be taken into account. We use the following notation: S^+ for the surface of the feed area, $C(\vec{r})$ for the middle axis of the printed strip, $C^+(\vec{r})$ for the middle axis of the feed points and $\vec{s}(\vec{r})$ for the direction of the tangent to C or C^+ respectively (Fig. 2.4). Then we can write the integral equation in the following form:

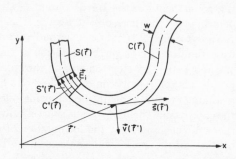

Fig. 2.4 Parameters for characterisation of a printed strip

$$\vec{s}(\vec{r}) \cdot \int\limits_S \overleftrightarrow{G}(\vec{r}, \vec{r}') \cdot \vec{J}_F(\vec{r}')dS' = \begin{cases} \vec{s}(\vec{r}) \cdot \vec{E}_i & , \vec{r} \in S^+ \\ 0 & , \vec{r} \in S \end{cases} \quad . \tag{2.18}$$

This is also based on the assumption of the feed areas being so small that their contribution to the radiation field can be neglected. For a conductive strip of length $2l$ parallel to the x-axis and fed with E_{ix} at the point $x = 0$ the following formulation is obtained:

$$\int\limits_{2l} \int\limits_w G_{xx}(x, y, x', y') J_{Fx}(x', y')dx'dy' = \begin{cases} E_{ix} & , \vec{r} \in S^+ \\ 0 & , \vec{r} \in S \end{cases} \quad . \tag{2.19}$$

The solution of the integral equation (2.18) and also (2.19) for all technically interesting cases is feasible only with the help of numerical methods. To be able to deal with large antennas in spite of limited computer storage, we shall assume that the transversal current distribution on the conductive strips can be determined by the equation of Fig. 2.5. The curve shown takes into account the singularity of the current at the edges of an infinitely thin, narrow conductive strip (see, for example, [2.4]) and it allows a simplification of the numerical analysis. The current $I_s(\vec{r})$ on the printed strips can be obtained by the integral

$$I_s(\vec{r})\,\vec{s}(\vec{r}) = \vec{I}(\vec{r}) = \int\limits_{-w/2}^{+w/2} \vec{J}_F(\vec{r}, v)dv \quad . \tag{2.20}$$

Thus, one only has to determine the current in the direction of the printed strips. For the final formulation of the integral equation it is now - as usual for the analysis of thin wire antennas ("Thin-wire-theory") - assumed that eqn. (2.18) has to be obeyed only on the middle axis of the conductive strips. Then

$$\vec{s}(\vec{r}) \cdot \int\limits_C \overleftrightarrow{G}(\vec{r}, \vec{r}') \cdot \vec{s}(\vec{r}')I_s(\vec{r}')ds' = \begin{cases} \vec{s}(\vec{r}) \cdot \vec{E}_i & , \vec{r} \in C^+ \\ 0 & , \vec{r} \in C \end{cases} \tag{2.21}$$

is obtained, with ds' being an element of length along the contour C. Because of condition (2.16) it is guaranteed that, when eqn. (2.21) holds, the tangential field distribution on the whole strip is approximately equal to zero.

As a comparison, the integral equations for the determination of the current distribution on a thin antenna in a homogeneous space which have so far been most important for antenna theory will be discussed briefly. The first formulation was by Pocklington in 1897 [2.5]. With the abbreviations

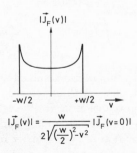

$$|\vec{J}_F(v)| = \frac{w}{2\sqrt{(\frac{w}{2})^2 - v^2}} |\vec{J}_F(v=0)|$$

Fig. 2.5 Transverse current distribution on a printed strip

$$g(x, x') = \frac{e^{-jkr_a}}{4\pi r_a j\omega\varepsilon} \tag{2.22}$$

and

$$r_a = \sqrt{(x - x')^2 + a^2}, \tag{2.23}$$

and a being the radius of the circular cylinder, the version used nowadays reads

$$\int_{-l}^{+l} I_x(x') \left(\frac{\partial^2 g(x, x')}{\partial x^2} + k^2 g(x, x') \right) dx' = \begin{cases} E_{ix} & ,x = 0 \\ 0 & ,0 < |x| \le l \end{cases} \tag{2.24}$$

In this equation the condition that the tangential electric field on the surface of the circular cylinder should be equal to zero is approximately substituted by the condition

$$E_x = 0 \tag{2.25}$$

on the cylinder axis. The term in brackets in the integrand is equal to $G_{xx}(x, y = 0, x', y' = a)$.

Having performed the differentiations in eqn. (2.24), one obtains the form which was chosen by Richmond as a starting point for the first numerical solution with the help of the method of moments [2.6]:

$$\int_{-l}^{+l} I_x(x') g(x, x') \frac{(1 + jkr_a)(2r_a^2 + a^2) + k^2 a^2 r_a^2}{r_a^4} dx' = \begin{cases} E_{ix} & ,x = 0 \\ 0 & ,0 < |x| \le l \end{cases} \tag{2.26}$$

The third important integral equation is that of Hallen ([2.7] and [2.8]):

$$\int_{-l}^{+l} I_x(x')g(x,x')dx' = -\frac{1}{k}(C_1\cos k\ x + \frac{U_1}{2}e^{-jk|x|}).$$ (2.27)

For its derivation it was assumed by Hallen that the antenna is fed in its centre through an infinitely thin gap with the voltage U_1 (in the literature often called "Delta-gap-excitation"). For the solution the constant C_1 has to be determined with the help of the boundary condition $I_x(x = \pm l) = 0$. Based on Hallen's integral equation a large number of approximate equations for the current on a thin cylindrical antenna have been developed (compare, for example, [2.9]). Eqn. (2.27), however, has the disadvantage that the form and the place of the feed are already fixed. Thus, because of its greater flexibility, eqn. (2.21) has been much more widely used for numerical methods. Here the determination of Green's function for free space can be performed without any difficulties. In contrast to this we first have to tackle the multi-layer problem (section 2.3) before the method of moments - described in appendix A1 - can be used for the solution of the integral equation (2.21).

All the integral equations mentioned so far describe an antenna problem characterized by the fact that the impressed sources are located on the printed circuit. If the sources are arranged in a finite or infinite distance from the conductive medium we call it a scattering problem. Integral equations for such a scattering problem can obviously be obtained by demanding that the electric field tangential to the surface of the scatterer is equal to zero. This field consists of the field of the external sources and that of the currents induced on the scatterer. In the case of a body consisting of a material of finite permittivity, after having introduced polarization currents, integral equations from the boundary conditions at the interface between the scatterer and its surroundings can be formulated. The solution of integral equations, formulated for scatterers embedded in plane stratified structures, often leads to similar difficulties to those of the antenna problem which is of interest here (see, for example, [2.10] - [2.13]).

2.3 Green's function of layered media

2.3.1 Summary of previous investigations

In 1909 Sommerfeld published an exact solution for the field of a vertical electric Hertzian dipole on a lossy homogeneous plane half-space in the form of an infinite integral [2.14]. The case of a horizontal Hertzian dipole was studied three years later by v. Hörschelmann [2.15]. In 1926 Sommerfeld published a summarizing paper, in which he also presented solutions for the dual problem, the magnetic point source [2.16]. Concerning the solutions for the infinite integrals, a large number of studies (as, for example, [2.17]) were published in the next 60 years. Investigation of all the cases which are interesting has only become possible now by the use of high speed computers, although in his study of 1909 Sommerfeld himself stated approximate solutions for two special cases. They describe the field of the Hertzian dipole as a superposition of space waves and a surface wave, which is in line with a solution of Maxwell's equations which Zenneck found some years before [2.18].

Sommerfeld based his work on solutions of the wave equation in cylindrical coordinates and thus obtained solutions consisting of integrals of Bessel functions. Weyl, however, in 1919, used Cartesian coordinates, which led to a double integral of exponential functions and which corresponds with a superposition of plane waves [2.19]. The solution of the double integral developed by Weyl did not seem to confirm the subdivision of the field into space and surface waves as proved by Sommerfeld; however, as early as 1920, Sommerfeld noted [2.20] that Weyl studied the case of field points located far away from the source and that then the surface wave can be neglected.

With the development of broadcasting in the 30's it became necessary to determine the propagation of electromagnetic waves depending on the electric properties of the earth surface more exactly than had been possible with the help of Sommerfeld's solution. This is only valid when the conductivity of the earth is very high and when displacement currents in the earth can be neglected. First, some studies of Niessen and Van der Pol ([2.21] to [2.27]) should be mentioned, in which an integral is derived without using cylindrical functions, which is different from the Sommerfeld approach, and where displacement currents are taken into account. Apart from this the source and observation point can be above the interface (in Sommerfeld's work the source lies within the interface). The result can be approximately interpreted with the help of an image source. Strutt [2.28] used Weyl's approach and in connection with reflection coefficients he obtained equations which are valid for field points at a greater distance from the ground. The approach of Wise ([2.29] and [2.30]) leads to power series which - although in a complicated way - can be used for observation points on the surface of the earth if the distance to the source is large. A generalization of Wise's result has been given by Rice [2.31]; because of the use of Legendre functions and hypergeometric functions, the formulas mentioned are very difficult to apply. In a study by Fock [2.32] equations for field points in the nearfield are presented, using cylindrical functions with an imaginary argument. To facilitate the design of antennas as well as of communication links, Norton - starting mainly from the results of Sommerfeld and Wise - deduced extensive equations for the fields of horizontal and vertical wire antennas with known current distribution, and he provided graphs, which make a simple evaluation of the equations possible ([2.33] to [2.36]).

As already mentioned, based on his calculations Weyl excluded the possibility of the field of a vertical current element on the ground being separable into space and surface wave. Also, the interpretations of the equations by Wise [2.37] and Norton [2.34] as well as measurements by Burrows [2.39] seemed to corroborate the fact that Zenneck's wave is not excited. According to the opinion of Epstein [2.40] a correct solution is obtained if the surface wave term is subtracted from the classical Sommerfeld solution. The same view is held by Kahan and Eckart (for example [2.41]), because, according to their opinion, Sommerfeld, when carrying out the integration, had overlooked one term, which precisely corresponds to the surface wave. In the following years, Bouwkamp especially ([2.42] and [2.43]), and Ott ([2.44] to [2.46]) as well as Kahan and Eckart ([2.47] and [2.48]) have commented on this topic. It turned out that, although Zenneck's wave is a part of the exact solution, unlike other terms it can be neglected for typical values of the complex dielectric constant of the earth when the observation point is in the farfield [2.49].

It was the aim of all the studies quoted so far to contribute to the design of communication links depending on the parameters of the earth. It was possible, however, to reverse the problem, which means to receive information about the structure of the earth beneath the surface by analysing the field of a point source on the ground. Therefore, it was necessary to give up visualizing the earth as a homogeneous lossy half-space and to base further studies on a stratified model. Early as 1933 [2.50] the case of a source above a structure consisting of two plane layers was discussed by means of extending the Sommerfeld approach. The discussion of the integrals was - (like [2.51] and [2.52]) - limited to the case where the displacement currents in the earth are negligible and where the field point is very close to the source. Based on the same two-layer model, Großkopf, working within the farfield of an antenna, tried to calculate the parameters of the layers ([2.53] to [2.55]) from the angle between the electric field vector and the surface of the ground. For this purpose he developed approximate equations starting with Weyl's approach. These equations correspond to those of Norton for the homogeneous half space [2.56]. Wait ([2.57] and [2.58]) systematically extended the Sommerfeld approach to a structure with an arbitrary number of layers. For that he started with the two required components of the Hertzian vector according to Sommerfeld with unknown amplitude coefficients in each layer; his equations correspond to those of a cascade of transmission lines. For the evaluation of the equations the stratified structure is described by means of a surface impedance, and thus, after solving the integrals according to the saddle-point method, one obtains a term which again corresponds to that of Norton for the homogeneous half-space.

The quasi-stationary solution for a magnetic Hertzian dipole on a two-layer structure has been discussed in 1955 [2.59]. Here especially the case investigated is one where the conductivity of the two layers varies only slightly. Curves for the components of the magnetic field on the ground - again based on the quasi-stationary solution - are given in a paper four years later [2.60]. An interesting case was examined by Wait in 1966 [2.61]: a layer consisting of low-loss material embedded in a layer with high losses. In this structure a wave can be guided as in a waveguide. In a subsequent paper Wait [2.62] discussed the case of a lossy layer on a layer of high conductivity. The complete field is described as a superposition of the field of the source itself, of an image source in free space and the contribution of an integral. This consists of the contributions of the poles and of the branch cuts; because of the fact that high losses have been assumed an approximate solution can be obtained. Formulas for the quasi-static case, i.e. for very small distances beween source and observation points are given for example by Weaver [2.63]. Within this range the image concept turned out to be a powerful approach for the solution of the Sommerfeld integrals. Up to now many papers have been published concerning the half-space problem including an extension of the theory beyond the quasi-static range (see, for example, [2.64] to [2.66]). An application of the image method to multi-layered media which are of interest here has already been given in the early 70's (for example [2.67] and [2.68]), where the quasi-static field over a two-layered lossy earth is discussed. The extension to the high-frequency range is investigated by Mahmoud et al. ([2.69] and [2.70]). They showed that the single image theory developed so far is adequate both at the quasi-static and the high frequency ends. An exact image theory valid for any frequency range was formulated in 1984 by Lindell and Alanen first for the half-space problem ([2.71]), recently for a grounded dielectric sheet ([2.72] and [2.73]). Starting with an idea first suggested in [2.66], the authors Laplace-transformed the reflection coefficient of the interface into a form which is Fourier transformable. Then an integral is obtained, which for example for a vertical magnetic dipole above a plane interface of two media can be interpreted as a field arising from a continuous image line source. For the grounded dielectric sheet the theory developed so far is applicable for low values of the dielectric constants of the substrate.

The formal extension of the Sommerfeld formulation towards the calculation of fields above layers of anisotropic materials has been developed in the years 1965 to 1969 ([2.74] to [2.77]). To solve the same problem Kong ([2.78]) described the field of each layer not - as had been common before - with the help of the Hertzian vector, but by using the superposition of a transversal-electric and a transversal-magnetic part. Because the two parts are not coupled by the boundary conditions these can easily be satisfied. General equations for the field of a source in a plane structure consisting of any number of - at first isotropic - layers are given in ([2.79] to [2.83]). In [2.84] these equations are formulated for the Cartesian components as a function of cylindrical coordinates which are found to be very handy in many cases. A paper of Sami and Samir [2.85] deals with the general case where the electric and magnetic properties of the layers must be described by means of permittivity and permeability tensors; further papers on this topic are [2.86] to [2.88].

The above mentioned papers ([2.74] to [2.88]) show that it is not particularly difficult to find general solutions based on Sommerfeld integrals even for complicated structures. The analysis of the integrals for concrete problems, however, demands a greater amount of work, because in most cases an extensive discussion of the analytical properties of the integrals has to be performed before one can use analytic or numerical methods. The saddle-point method has proved to be particularly useful (see for example [2.89]). Thus, with this method, the Sommerfeld solution is obtained in a much easier way than that in which it was derived by Sommerfeld himself. Therefore, in the following, some papers are mentioned in which the integrals are solved by the saddle-point method: Papousek and Schnizer ([2.90] and [2.91]) investigate the radiation field of a source in the three-layer medium (earth-air-ionosphere). The excitation of surface waves on a dielectric sheet by a source above the sheet is determined by Gütl et al. ([2.92]). Papers of Kong and others ([2.93] to [2.99]) extensively discuss the case of a source above a geological structure consisting of two layers. Here first of all the singularities of the integrands - poles and branch points - are determined, then the contributions which are obtained with the help of the saddle-point method are derived and interpreted physically.

Apart from the papers quoted so far, extensive studies of the complete field of interest have been published in monographs. A summary of Sommerfeld's investigations can be found in the book "Partielle Differentialgleichungen der Physik" [2.100]. In order to derive the equations and diagrams for the induction method, Keller and Frischknecht [2.101] discuss in their book solutions that neglect displacement currents. Wait's monograph [2.102] starts with the reflection and transmission coefficients of waves with oblique angles of incidence. The solution for the field in each layer is given by using the Hertzian vector. Felsen's book [2.89] contains a comprehensive explanation of the saddle-point method and an extensive discussion of the field of a point source above a lossy dielectric half space as well as above a grounded dielectric layer. The book of King and Smith [2.103] is mainly concerned with wire antennas (dipoles, loops), and additionally gives plenty of information about technical problems. The concept of Kong [2.104] is based on the TM/TE-waves approach.

To solve our problem the field of a horizontal electric Hertzian dipole in a stratified medium must be known, as has been explained in section 2.2. In 2.3.2 the Hertzian vector approach, as it has been applied by Sommerfeld for the field of a source on a homogeneous lossy half-space, is briefly illustrated.

For a multi-layer structure the representation of the field using TE and TM waves proves to be very useful. Therefore, in 2.3.3, from solutions of the wave equation in cylindrical coordinates, general equations of the field components with still unknown amplitude coefficients are formulated. In the appendices A2 and A3 it is shown that by satisfying the boundary conditions all amplitude coefficients can be determined. Here the equations are prepared in such a way that a numerical analysis can easily be carried out. To achieve a clearly arranged display, at the end of appendix A3 the equations for the field components in all layers are given in tables. Methods for the solution of integrals are described in section 2.4. A brief dicussion of some analytical properties of the integrands in 2.4.1 is followed by the discussion of two special cases:

A) For the determination of the farfield of an antenna with equation (2.14) Green's function has to be determined for field points which are far away from the source. Particularly suitable for this is the saddle-point method described in section 2.4.2. The application of this method for a structure with many layers is schematically possible; the results achieved often allow a simple physical interpretation.

B) For the solution of the integral equation (2.18) the field of a point source in observation points in the nearfield of the source has to be determined. In this case usually only a numerical integration (Section 2.4.3) leads to useful results. If the source and observation points lie very close to each other approximate solutions can be used. They are developed in section 2.4.4.

2.3.2 The Hertzian vector approach

Sommerfeld's starting point is the two Hertzian vectors $\vec{\Pi}_e$ and $\vec{\Pi}_m$ that are defined by

$$\vec{H} = j\omega\varepsilon \, \text{rot}\vec{\Pi}_e \tag{2.28}$$

for the electric source currents, and by

$$\vec{E} = -j\omega\mu \, \text{rot}\vec{\Pi}_m \tag{2.29}$$

for the magnetic source currents. The antennas this study focuses on will be described solely by means of electric source currents. Therefore, only the solution for $\vec{\Pi}_e$ will be discussed below, to facilitate writing

$$\vec{\Pi}_e = \vec{\Pi} \quad . \tag{2.30}$$

This Hertzian vector must obey the wave equation

$$\Delta\vec{\Pi} + k_i^2\vec{\Pi} = 0 \tag{2.31}$$

within the source–free regions of the i-th layer. From the boundary conditions it follows for the vertical dipole that $\vec{\Pi}$ has only one component in the direction of the current (Fig. 2.6a):

$$\vec{\Pi} = \vec{u}_z\Pi_z \quad . \tag{2.32}$$

Theory

In the case of the horizontal current element (Fig. 2.6b) a component perpendicular to the surface of the earth has to be added to the Hertzian vector:

$$\vec{\Pi} = \vec{u}_x \Pi_x + \vec{u}_z \Pi_z \quad . \tag{2.33}$$

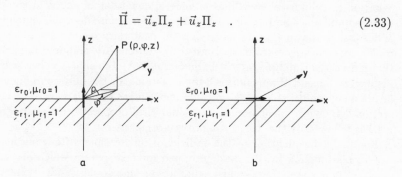

Fig. 2.6 **Vertical (a) and horizontal (b) Hertzian dipole in the interface between two different dielectrics**

Sommerfeld developed a suitable approach for the components of the Hertzian vector through the general solution of the wave equation in cylindrical coordinates for regions that include the z-axis. It reads

$$\Pi_{x,z} = C_{x,z} \, J_{n_\varphi}(k_\rho \rho) \, e^{\pm j k_{zi} z} F_{n_\varphi}(\varphi) \tag{2.34}$$

with

$$F_{n_\varphi}(\varphi) = \begin{cases} \sin(n_\varphi \varphi) \\ \cos(n_\varphi \varphi) \end{cases} \quad . \tag{2.35}$$

$$k_{zi} = \sqrt{k_i^2 - k_\rho^2} \tag{2.36}$$

is the condition of separability and

$$n_\varphi = 0, 1, 2, 3, \ldots \tag{2.37}$$

holds in order to achieve uniqueness. The constant C is determined by the dipole moment, the frequency and the electric properties of the space. The vector potential of a Hertzian dipole in free space can be formulated with the help of the Fourier-Bessel representation as a superposition of solutions of the wave equation in cylindrical coordinates:

$$\frac{e^{-jkr}}{r} = -j \int\limits_0^\infty \frac{k_\rho dk_\rho}{\sqrt{k^2 - k_\rho^2}} \, J_0(k_\rho \rho) \, e^{\mp j \sqrt{k^2 - k_\rho^2} \, z} \quad . \tag{2.38}$$

The negative sign in the exponent is valid for $z > 0$ and the positive one for $z < 0$. Sommerfeld combines a "primary" and a "secondary" excitation to achieve the complete solution. The first one corresponds to the field of a source in a homogeneous space with the parameters $\varepsilon_0\varepsilon_{r0}$ and $\mu_0\mu_{r0}$ or $\varepsilon_0\varepsilon_{r1}$ and $\mu_0\mu_{r1}$ respectively, the second one takes into account the inhomogeneous medium. Thus one obtains for the vertical dipole:

$$\Pi_{z0/1} = C_{0/1}\frac{e^{-jk_{0/1}r}}{r} + \int\limits_0^\infty f_{0/1}(k_\rho)\,J_0(k_\rho\rho)\,e^{\mp j\sqrt{k_{0/1}^2 - k_\rho^2}\,z}\,dk_\rho \quad. \quad (2.39)$$

The first figures of the subscripts separated by a slash belong to the equations for region "0", the second to those for region "1". The functions $f_0(k_\rho)$ and $f_1(k_\rho)$ are determined in such a way that the boundary conditions are fulfilled at the interfaces. To achieve this, the equations of the field components in both regions are established using Maxwell's equations and the tangential field components are set equal in the plane $z = 0$. Then for $\Pi_{z0/1}$ one gets the equations

$$k_0^2\Pi_{z0} = k_1^2\Pi_{z1}\,, \quad z = 0 \quad\quad\quad\quad\quad\quad (2.40)$$

and

$$\frac{\partial\Pi_{z0}}{\partial z} = \frac{\partial\Pi_{z1}}{\partial z}\,, \quad z = 0 \quad. \quad\quad\quad\quad (2.41)$$

The insertion of eqn. (2.39) into the eqns. (2.40) and (2.41) leads to

$$\Pi_{z0/1} = C_{0/1}\int\limits_0^\infty -j\frac{2k_{1/0}^2}{N_1}\,k_\rho J_0(k_\rho\rho)\,e^{\mp j\sqrt{k_{0/1}^2 - k_\rho^2}\,z}\,dk_\rho \quad\quad (2.42)$$

with

$$N_1 = k_0^2\sqrt{k_1^2 - k_\rho^2} + k_1^2\sqrt{k_0^2 - k_\rho^2} \quad. \quad\quad\quad (2.43)$$

Again, the upper sign in the exponent is valid for $z > 0$ and the lower sign for $z < 0$.

In the case of the horizontal dipole, additional equations have to be formulated for the x-component of the Hertzian vector, where care has to be taken of the boundary conditions. For the components of the Hertzian vectors they read:

$$k_0^2\Pi_{x0} = k_1^2\Pi_{x1} \quad, \quad\quad\quad\quad\quad\quad\quad (2.44)$$

$$k_0^2\frac{\partial\Pi_{x0}}{\partial z} = k_1^2\frac{\partial\Pi_{x1}}{\partial z} \quad, \quad\quad\quad\quad\quad (2.45)$$

$$k_0^2\Pi_{z0} = k_1^2\Pi_{z0} \quad, \quad\quad\quad\quad\quad\quad (2.46)$$

$$\frac{\partial\Pi_{z0}}{\partial z} - \frac{\partial\Pi_{z1}}{\partial z} = \frac{\partial\Pi_{x1}}{\partial x} - \frac{\partial\Pi_{x0}}{\partial x} \quad. \quad\quad\quad (2.47)$$

Obviously the components Π_{x0} and Π_{x1} are coupled by the boundary conditions. With a formulation similar to that for the vertical dipole the final result is:

$$\Pi_{x0/1} = C_{0/1} \frac{k_0^2}{k_{0/1}^2} \int_0^\infty -2j \frac{J_0(k_\rho\rho)}{N_2} e^{\mp j \sqrt{k_{0/1}^2 - k_\rho^2}\, z}\, k_\rho\, dk_\rho \quad , \quad (2.48)$$

$$\Pi_{z0/1} = C_{0/1}(k_0^2 - k_1^2)\frac{k_0^2}{k_{0/1}^2}\cos\varphi \tag{2.49}$$

$$\int_0^\infty -2 \frac{J_0'(k_\rho\rho)}{N_1 N_2} e^{\mp j \sqrt{k_{0/1}^2 - k_\rho^2}\, z}\, k_\rho^2\, dk_\rho$$

with

$$N_2 = \sqrt{k_0^2 - k_\rho^2} + \sqrt{k_1^2 - k_\rho^2} \tag{2.50}$$

and N_1 taken from eqn. (2.43).

2.3.3 TM- and TE-waves approach for a horizontal electric point source

Sommerfeld's approach has the disadvantage that for the horizontal point source the contributions of $\Pi_{x0/1}$ and $\Pi_{z0/1}$ have to be considered together in order to fulfil the boundary conditions at the interfaces. This disadvantage can be avoided if the complete field is subdivided into a part which is transverse magnetic (TM), and a part which is transverse electric (TE) with respect to the z-coordinate [2.78]. The TM as well as the TE field obeys the boundary conditions at the interfaces; both parts are coupled only by the source. This property makes the representation considerably more simple, especially in the case of media with many layers. Therefore, in this study the field will be described with the help of TM and TE fields.

It is a well known fact that TM and TE fields with respect to z are determined by a setup of two vector potentials which only have a z-component:

$$\vec{H}^{TM} = \text{rot}\vec{A} = \text{rot}(\vec{u}_z \Psi^{TM}) \quad , \tag{2.51}$$

$$\vec{E}^{TE} = -\text{rot}\vec{F} = \text{rot}(\vec{u}_z \Psi^{TE}) \quad . \tag{2.52}$$

Substituting Ψ^{TM} and Ψ^{TE} in Maxwell's equations leads to the eqn.

$$\Delta\Psi^{TM/TE} + k_i^2\, \Psi^{TM/TE} = 0 \quad . \tag{2.53}$$

(parameters of the media $\mu_0\mu_{ri}, \varepsilon_0\varepsilon_{ri}$). Thus, we have retrieved the scalar wave equation as in Sommerfeld's approach. The solutions for the field can be developed in an analogous way:

$$\Psi_{n_\varphi,k_i} = \int_0^\infty \left[A(k_\rho)e^{+jk_{zi}z} + B(k_\rho)e^{-jk_{zi}z} \right] J_{n_\varphi}(k_\rho\rho) \begin{Bmatrix} \sin(n_\varphi\varphi) \\ \cos(n_\varphi\varphi) \end{Bmatrix} dk_\rho .$$

$$\tag{2.54}$$

The amplitudes $A(k_\rho)$ and $B(k_\rho)$ are determined by the source and by the boundary conditions at the interfaces $z = $ const.

A more suitable representation for the analysis is obtained if the Bessel function is replaced by the Hankel function. Then, one obtains an integral within the limits $-\infty$ and $+\infty$ [2.100]:

$$\Psi_{n_\varphi, k_i} = \frac{1}{2} \int_{-\infty}^{\infty} \left[A(k_\rho)e^{+jk_{zi}z} + B(k_\rho)e^{-jk_{zi}z} \right] H_{n_\varphi}^{(2)}(k_\rho\rho) \begin{Bmatrix} \sin(n_\varphi\varphi) \\ \cos(n_\varphi\varphi) \end{Bmatrix} dk_\rho \,.$$

(2.55)

The path of integration of this complex integral (2.55) has to be chosen suitably. The equations are a little bit simpler, if the components of the field in the z-direction, E_z^{TM} and H_z^{TE}, are chosen as a starting point instead of Ψ^{TM} and Ψ^{TE} [2.78]. Then the equations for the transverse components read

$$\vec{E}_t^{TM}(k_\rho) = \frac{1}{k_\rho^2} \vec{\nabla}_t \frac{\partial E_z^{TM}(k_\rho)}{\partial z} \quad , \tag{2.56}$$

$$\vec{E}_t^{TE}(k_\rho) = -\frac{j\omega\mu_0\mu_{ri}}{k_\rho^2} \vec{\nabla}_t \times [\vec{u}_z H_z^{TE}(k_\rho)] \quad , \tag{2.57}$$

$$\vec{H}_t^{TM}(k_\rho) = \frac{j\omega\varepsilon_0\varepsilon_{ri}}{k_\rho^2} \vec{\nabla}_t \times [\vec{u}_z E_z^{TM}(k_\rho)] \quad , \tag{2.58}$$

$$\vec{H}_t^{TE}(k_\rho) = \frac{1}{k_\rho^2} \vec{\nabla}_t \frac{\partial H_z^{TE}(k_\rho)}{\partial z} \tag{2.59}$$

with

$$\vec{\nabla}_t \, a = \left(\frac{\partial}{\partial\rho} \, \vec{u}_\rho + \frac{1}{\rho} \frac{\partial}{\partial\varphi} \, \vec{u}_\varphi \right) a \tag{2.60}$$

and

$$\vec{\nabla}_t \times (a \, \vec{u}_z) = \left(\frac{1}{\rho} \frac{\partial}{\partial\varphi} \, \vec{u}_\rho - \frac{\partial}{\partial\rho} \, \vec{u}_\varphi \right) a \quad . \tag{2.61}$$

Thus, the equations for all field components, are ($F_{n_\varphi}^{TM/TE}$ is used as an abbreviation for the φ -dependence):

$$E_\rho^{TM} = \int_{-\infty}^{\infty} -j\frac{k_{zi}}{k_\rho} \left\{ -A(k_\rho)e^{+jk_{zi}z} + B(k_\rho)e^{-jk_{zi}z} \right\}$$

$$H_{n_\varphi}^{(2)'}(k_\rho\rho) \, F_{n_\varphi}^{TM}(\varphi) \, dk_\rho \,, \tag{2.62}$$

$$E_\varphi^{TM} = \int_{-\infty}^{\infty} -j\frac{k_{zi}}{k_\rho^2\rho} \left\{ -A(k_\rho)e^{+jk_{zi}z} + B(k_\rho)e^{-jk_{zi}z} \right\}$$

$$H_{n_\varphi}^{(2)}(k_\rho\rho) \, F_{n_\varphi}^{TM'}(\varphi) \, dk_\rho \,, \tag{2.63}$$

$$E_z^{TM} = \int\limits_{-\infty}^{\infty} \left\{ A(k_\rho)e^{+jk_{zi}z} + B(k_\rho)e^{-jk_{zi}z} \right\}$$

$$H_{n_\varphi}^{(2)}(k_\rho\rho)\, F_{n_\varphi}^{TM}(\varphi)\, dk_\rho\,, \qquad (2.64)$$

$$H_\rho^{TM} = \int\limits_{-\infty}^{\infty} j\frac{\omega\varepsilon_0\varepsilon_{ri}}{k_\rho^2\rho} \left\{ A(k_\rho)e^{+jk_{zi}z} + B(k_\rho)e^{-jk_{zi}z} \right\}$$

$$H_{n_\varphi}^{(2)}(k_\rho\rho)\, F_{n_\varphi}^{TM\prime}(\varphi)\, dk_\rho\,, \quad (2.65)$$

$$H_\varphi^{TM} = \int\limits_{-\infty}^{\infty} -j\frac{\omega\varepsilon_0\varepsilon_{ri}}{k_\rho} \left\{ A(k_\rho)e^{+jk_{zi}z} + B(k_\rho)e^{-jk_{zi}z} \right\}$$

$$H_{n_\varphi}^{(2)\prime}(k_\rho\rho)\, F_{n_\varphi}^{TM}(\varphi)\, dk_\rho\,, \quad (2.66)$$

$$H_z^{TM} = 0 \,, \qquad\qquad (2.67)$$

$$E_\rho^{TE} = \int\limits_{-\infty}^{\infty} -j\frac{\omega\mu_0\mu_{ri}}{k_\rho^2\rho} \left\{ C(k_\rho)e^{+jk_{zi}z} + D(k_\rho)e^{-jk_{zi}z} \right\}$$

$$H_{n_\varphi}^{(2)}(k_\rho\rho)\, F_{n_\varphi}^{TE\prime}(\varphi)\, dk_\rho\,, \quad (2.68)$$

$$E_\varphi^{TE} = \int\limits_{-\infty}^{\infty} j\frac{\omega\mu_0\mu_{ri}}{k_\rho} \left\{ C(k_\rho)e^{+jk_{zi}z} + D(k_\rho)e^{-jk_{zi}z} \right\}$$

$$H_{n_\varphi}^{(2)\prime}(k_\rho\rho)\, F_{n_\varphi}^{TE}(\varphi)\, dk_\rho\,, \quad (2.69)$$

$$E_z^{TE} = 0 \,, \qquad\qquad (2.70)$$

$$H_\rho^{TE} = \int\limits_{-\infty}^{\infty} -j\frac{k_{zi}}{k_\rho} \left\{ -C(k_\rho)e^{+jk_{zi}z} + D(k_\rho)e^{-jk_{zi}z} \right\}$$

$$H_{n_\varphi}^{(2)\prime}(k_\rho\rho)\, F_{n_\varphi}^{TE}(\varphi)\, dk_\rho\,, \quad (2.71)$$

$$H_\varphi^{TE} = \int\limits_{-\infty}^{\infty} -j\frac{k_{zi}}{k_\rho^2\rho} \left\{ -C(k_\rho)e^{+jk_{zi}z} + D(k_\rho)e^{-jk_{zi}z} \right\}$$

$$H_{n_\varphi}^{(2)}(k_\rho\rho)\, F_{n_\varphi}^{TE\prime}(\varphi)\, dk_\rho\,, \quad (2.72)$$

$$H_z^{TE} = \int\limits_{-\infty}^{\infty} \left\{ C(k_\rho)e^{+jk_{zi}z} + D(k_\rho)e^{-jk_{zi}z} \right\} H_{n_\varphi}^{(2)}(k_\rho\rho)\, F_{n_\varphi}^{TE}(\varphi)\, dk_\rho\,. \quad (2.73)$$

This approach can be interpreted in such a way that the complete field consists of a superposition of inhomogeneous cylindrical waves with the wave amplitudes A, B, C and D ([2.105]). The determination of these amplitudes is carried out in the appendices A2 and A3.

2.4 Solution of the Sommerfeld integral

2.4.1 Some analytical properties of the integrands

In this section some general properties of the integrands, which are important for the application of analytical and numerical methods, will be discussed briefly. It is obviously sufficient to analyse the integrands of the equations for the field components dE_{zi}^{TM} and dH_{zi}^{TE}. Further details will be given in connection with the discussion of special applications.

Eqn. (2.36) connects the wavenumbers k_{zi} with the integration variable

$$k_{zi} = \sqrt{k_i^2 - k_\rho^2} \quad . \tag{2.74}$$

The double-valued square root is made unique by the following physical considerations: in the ranges $\hat{m}$ and n (Fig. 2.1) the dependence on z is described by $e^{-jk_{z\hat{m}}z}$ and $e^{+jk_{zn}z}$ respectively. As only outgoing waves are allowed to exist,

$$Re\{k_{z\hat{m}/n}\} > 0 \tag{2.75}$$

must hold. To ensure that the field vanishes for $|z| \to \infty$, which is a further requirement of the radiation condition, we must have

$$Im\{k_{z\hat{m}/n}\} < 0 \quad . \tag{2.76}$$

The last condition is identical with the requirement that the integrands remain bounded for $|z| \to \infty$.

Because of the fact that the integrands are even functions of k_{zi} ($i \neq \hat{m}, n$) - as can be shown easily ([2.89], p. 456) - the integrands are regular at the points $k_{zi} = 0$ ($i \neq \hat{m}, n$). Thus, if the Bessel function (for example eqn. (2.54)) is used, the integrands have only branch points at $k_{\rho b1} = \pm k_{\hat{m}}$ and $k_{\rho b2} = \pm k_n$. With the Hankel function, for example, in eqns. (A3.47) to (A3.58) another branch point occurs at $k_\rho = 0$. In the case of Fig. 2.7 we have $|k_{\hat{1}}| \to \infty$ because of $\kappa_{\hat{1}} \to \infty$. Thus, in that case only the branch points at $k_{\rho b2} = \pm k_n$ and, should the occasion arise, at $k_\rho = 0$ have to be taken into account.

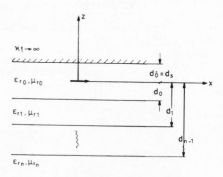

Fig. 2.7 Stratified medium with reflector at $z = d_s$

Apart from the branch points the integrands generally possess simple poles. For the analysis of the integrals it has to be checked whether these poles lie on the allowed or disallowed sheet of the Riemann surface. As will be shown later, also poles on the disallowed sheet can contribute to the solution.

2.4.2 Solution by the method of steepest descent

To simplify the discussion of the method of steepest descent, this section is based on the equations valid for a structure with a semiinfinite region "0" ($d_0^- \to \infty$ in Fig. 2.7). Then the integrals which have to be solved can be written generally in the form

$$I = \int\limits_{-\infty}^{+\infty} g(k_\rho)\, e^{-j k_{z0} z'}\, H_1^{(2)}(k_\rho \rho)\, dk_\rho \quad . \tag{2.77}$$

The meaning of the abbreviations $g(k_\rho)$ and z' can be deduced easily from the eqns. (A3.47) to (A3.58). For larger distances from the source ($\rho \gg \lambda$) the Hankel function can be approximately substituted by the exponential function

$$H_1^{(2)}(x) \approx \sqrt{\frac{2}{\pi x}}\, e^{-j\left(x - \frac{3\pi}{4}\right)} \quad . \tag{2.78}$$

The starting point for the method of steepest descent — another common name is the "saddle–point method"— is the transformation ([2.89], p. 462)

$$k_\rho = k_0 \sin w \quad , \tag{2.79}$$

from which the equation for k_{z0}

$$k_{z0} = \sqrt{k_0^2 - k_\rho^2} = k_0 \cos w \tag{2.80}$$

follows. Using spherical coordinates

$$\rho = r' \sin\vartheta \quad , \tag{2.81}$$

$$z' = r' \cos\vartheta \tag{2.82}$$

and

$$r' = \sqrt{\rho^2 + z'^2} \tag{2.83}$$

the following transformed form is obtained for the integral (2.77)

$$I = \int\limits_{W} j \sqrt{\frac{2}{\pi k_0 r' \sin w \, \sin\vartheta}}\; e^{j\frac{\pi}{4}}\, g(k_0 \sin w\,)\, e^{-j k_0 r' \cos(w - \vartheta)}\, k_0 \cos w \; dw \, .$$

$$\tag{2.84}$$

Applying eqn. (2.79) the complex k_ρ-plane is mapped onto the complex w-plane. By doing this, the two sheets of the Riemann surface which exist because the branch points at $\pm k_0$ are mapped onto strips of width π. Thus, the transformation removes the branch points at $\pm k_0$.

Fig. 2.8 shows the areas corresponding to each other. For this it has been assumed that k_0 is real, the branch cuts extend on the real axis towards $\pm\infty$ and that $Re\{k_{z0}\} > 0$ holds on the upper sheet. As an abbreviation the real and the imaginary part of a complex number will be marked by a single and a double prime, respectively. Thus, for example, for k_{z0} one gets

$$k_{z0} = k'_{z0} + j k''_{z0} \quad . \tag{2.85}$$

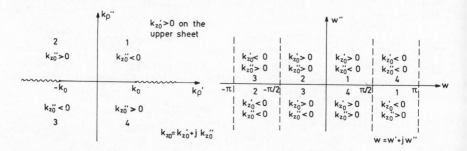

Fig. 2.8 Mapping of the k_ρ-onto the w-plane by the transformation $k_\rho = k_0 \sin w$

As a simplification for what follows, eqn. (2.84) is written in the form

$$I = \int_W f(w)\, e^{R\, q(w)}\, dw \tag{2.86}$$

with

$$q(w) = -j \cos(w - \vartheta) \quad . \tag{2.87}$$

Because $\rho \gg \lambda_0$, and thus $r' \gg \lambda_0$,

$$k_0 r' = R \gg 1 \tag{2.88}$$

holds in this equation (k_0 being real).

For the solution of the integral the path of the integration is chosen in such a way that it runs through the "saddle point" w_s. The characteristic feature of this point is that $Re\{q(w)\}$ has a maximum value at w_s and that $Re\{q(w)\} \leq Re\{q(w_s)\}$ on the rest of the path. The function $e^{Rq(w)}$ has a maximum at w_s, too. Because R is a large number, it is sufficient for an approximate solution of the integral to analyse the integral only in the vicinity of w_s. Based on the assumption that $f(w)$ is regular and slowly varying in the vicinity of w_s, one can set $f(w) \approx f(w_s)$ and take it outside the integral. For the solution of the integral, $q(w)$ is expanded as a power series round w_s, the first few terms of which are taken into account.

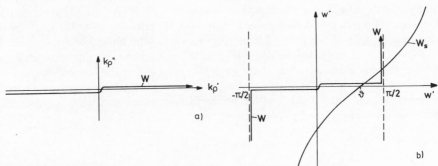

Fig. 2.9   Path of integration in the k_ρ- and the
w-plane

W: path along the real k_ρ- axis
W_s: path of steepest descent

According to [2.89] the location of the saddle point can be determined from the equation

$$\frac{d\,q(w)}{dw} = 0 \quad . \tag{2.89}$$

In our case

$$\frac{d\cos(w - \vartheta)}{dw} = 0 \tag{2.90}$$

leads to the solution

$$w_s = \vartheta \quad . \tag{2.91}$$

Thus, for all field points in the farfield the saddle point lies in the range

$$0 < Re\{w_s\} < \tfrac{\pi}{2} \tag{2.92}$$

on the real axis.

The path of steepest descent, here called W_s, is where, starting with w_s, the magnitude of $e^{+Rq(w)}$, and thus also that of $Re\{q(w)\}$, decreases most rapidly. If an element of length along the path of integration through w_s is called ds, then $\frac{dRe\{q(w)\}}{ds}$ must assume a maximum. From this, with the help of Cauchy-Riemann's differential equations, it follows that the condition stated is fulfilled for

$$Im\{q(w)\} = \text{const} \quad . \tag{2.93}$$

Thus, for our case, using eqn. (2.87), the path of steepest descent is defined by the equation

$$Im\{-j\cos(w - \vartheta)\} = -1 \quad . \tag{2.94}$$

As shown in Fig. 2.9 it intersects the real w-axis in the saddle point at an angle of 45°. Also, the original path of integration W is plotted. According to Cauchy's theorem the deformation of the path W into the path W_s does not change the value of the integral if no singularities of the integrand are crossed; otherwise the singularities between W_s and W must be taken into account by adding the contributions of the residues of the poles and of the contour integrals for the branch cuts. From eqn. (2.94) it can be seen which singularities lie between W and W_s. If the location of the branch points and poles is denoted with the subscripts b and p, respectively,

$$w_{b,p} = w'_{b,p} + j w''_{b,p} \tag{2.95}$$

one concludes using eqn. (2.94) that

$$\vartheta \geq \vartheta_{b,p} = w'_{b,p} - \mathrm{sign} w''_{b,p} \ \arccos(\frac{1}{\cosh w''_{b,p}}) \tag{2.96}$$

must hold, to make sure that the singularity in $w_{b,p}$ lies between W and W_s.

Thus, the solution of the integral (2.86) consists of the contribution of the saddle point, the poles and the branch cuts:

$$I = I_s + I_p + I_b \quad . \tag{2.97}$$

Here the solution is given for the case of the poles and the branch points not lying near the saddle point, so that their contributions can be calculated separately. One obtains ([2.89]):

$$I = \int_W f(w) e^{-j k_0 r' \cos(w - \vartheta)} \, dw$$

$$\approx \sqrt{\frac{2\pi}{k_0 r'}} \ f(\vartheta) e^{-j(k_0 r' - \frac{\pi}{4})}$$

$$+ U(\vartheta - \vartheta_p)(-2\pi j) \left[(w - w_p) f(w) \right]_{w_p} e^{-j k_0 r' \cos(w_p - \vartheta)} \tag{2.98}$$

$$+ \frac{2\sqrt{\pi}}{|k_0 r' \sin(\vartheta - w_b)|^{\frac{3}{2}}} \left[\sqrt{w - w_b} \ \frac{d f(w)}{dw} \right]_{w_b}$$

$$\cdot e^{-j k_0 r' \cos(w_b - \vartheta)} + j \frac{3}{2} \left(\frac{\pi}{2} + \arg(k_0 \sin(w_b - \vartheta)) \right) \quad .$$

As already mentioned, for the derivation of the first term on the right-hand side of eqn. (2.98) the $q(w)$ are expanded into a power series, of which the first three

terms are taken into account. The second term of eqn. (2.98) is due to the
residue theorem. The function $U(x)$ is equal to 1 for $x > 0$ and vanishes for
$x < 0$. The derivation of the third term of eqn. (2.98) requires a discussion of
the function $f(w)$ in the vicinity of the branch point. As such a discussion in
general is very difficult we will not go into further detail here.

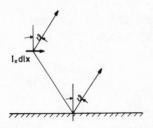

Fig. 2.10 Geometric-optical interpretation of the saddle-point solution

Using eqn. (2.98) the contribution of the saddle point is easily found for
structures with a semifinite region "0" ($d_{\hat{0}} \to \infty$ in Fig. 2.7) from the eqns.
(A3.47) to (A3.58). Thus, with eqn. (2.78)

$$H_1^{(2)'}(x) \approx -jH_1^{(2)}(x) \quad . \tag{2.99}$$

One obtains for the components of the $E^{TM/TE}$ field, marked by the additional
subscript "s":

$$dE_{\rho s}^{TM} = -j\frac{I_x dl_x k_0^2}{4\pi\omega\varepsilon_0\varepsilon_{r0}}(1 - \Gamma_0^{TM}(\vartheta)e^{-2jk_0 d_0 \cos\vartheta})\frac{e^{-jk_0 r}}{r}\cos^2\vartheta\cos\varphi \quad ,(2.100)$$

$$dE_{\varphi s}^{TM} = -\frac{I_x dl_x}{4\pi\omega\varepsilon_{r0}}\frac{k_0}{\varepsilon_0}(1 - \Gamma_0^{TM}(\vartheta)e^{-2jk_0 d_0 \cos\vartheta})\frac{1}{\rho}\frac{e^{-jk_0 r}}{r}\frac{\cos^2\vartheta}{\sin\vartheta}\sin\varphi \quad ,(2.101)$$

$$dE_{zs}^{TM} = j\frac{I_x dl_x k_0^2}{4\pi\omega\varepsilon_0\varepsilon_{r0}}(1 - \Gamma_0^{TM}(\vartheta)e^{-2jk_0 d_0 \cos\vartheta})\frac{e^{-jk_0 r}}{r}\cos\vartheta\sin\vartheta\cos\varphi \quad ,(2.102)$$

$$dE_{\rho s}^{TE} = \frac{I_x dl_x}{4\pi}\frac{\omega\mu_0\mu_{r0}}{k_0}(1 + \Gamma_0^{TE}(\vartheta)e^{-2jk_0 d_0 \cos\vartheta})\frac{1}{\rho}\frac{e^{-jk_0 r}}{r}\frac{\cos\varphi}{\sin\vartheta} \quad , \tag{2.103}$$

$$dE_{\varphi s}^{TE} = j\frac{I_x dl_x}{4\pi}\omega\mu_0\mu_{r0}(1 + \Gamma_0^{TE}(\vartheta)e^{-2jk_0 d_0 \cos\vartheta})\frac{e^{-jk_0 r}}{r}\sin\varphi \quad . \tag{2.104}$$

The field components $E_{\rho s}^{TM}$, E_{zs}^{TM} and $E_{\varphi s}^{TE}$ are proportional to $1/r$, $E_{\varphi s}^{TM}$ and
$E_{\rho s}^{TE}$ are proportional to $1/(\rho r)$; thus, provided that $\rho \gg \lambda$, the latter can be
neglected. The ρ- and the z-component can be combined in order to obtain the

ϑ-component:

$$dE_{\vartheta s}^{TM} = dE_{\rho s}^{TM}\cos\vartheta \;-\; dE_{zs}^{TM}\sin\vartheta$$

$$= -j\frac{I_x dl_x}{4\pi\omega\varepsilon_0\varepsilon_{r0}}k_0^2 \left(1 - \Gamma_0^{TM}(\vartheta)\,e^{-2jk_0 d_0\cos\vartheta}\right)\frac{e^{-jk_0 r}}{r}\cos\vartheta\,\cos\varphi\;.$$

$$(2.105)$$

It follows from the eqns. (2.100) to (2.105) that the contribution of the saddle point gives the solution, which also can be deduced by arguments of geometrical optics (Fig. 2.10).

2.4.3 Numerical solution

2.4.3.1 General comments

As already mentioned, the Sommerfeld integrals have to be solved numerically for the distances between source and observation points in the range of 0.1λ to 3.0λ. A computer program also offers the possibility of checking the range of validity of the approximate solutions. The application of numerical methods of integration has been made possible by the development of fast computers. Most of the papers published so far deal with the problem of a source above a lossy homogeneous half-space, for example, [2.17]. A fairly obvious procedure, i.e. the direct numerical integration of, for example, the eqns. (2.42), (2.48) and (2.49) along the k_ρ-axis, is performed in [2.106] and [2.107]. If the inequality $\rho < z + d_0$ holds for observation points $P(\rho,\varphi,z)$ above the ground (d_0 is the height of the source point above the ground), the integrand strongly oscillates near the branch points. In [2.108] it is shown that this oscillation can be removed by an adequate transformation. The shift of the path of integration into the third and fourth quadrant of the k_ρ-plane is performed, for example, in [2.109] and [2.110], because then, due to the asymptotic behaviour of the cylindrical functions, a fast convergence is achieved. It is very disadvantageous, however, that in this case cylindrical functions with complex arguments have to be evaluated. In [2.111] to [2.113] the integration is performed along the path of steepest descent. This results in a very good convergence of the integrals which have to be analysed, though the calculation of the contributions of poles and branch cuts is laborious. For the application of numerical methods of integration on multi-layer problems, see [2.94]. Again, the real k_ρ-axis was chosen as the path of integration. As the layers were assumed to be lossy, the path of the integration does not touch any pole singularity.

When investigating microstrip antennas, the poles lie very close below the real k_ρ-axis because of the low–loss substrate. Then the contribution of these poles can be determined with the help of the residue theorem [2.114]. (A detailed discussion of this case will be given in section 3.2.4.) The shift of the path of integration into the complex k_ρ-plane in the case of microstrip antennas is discussed in [2.115]. In order to be able to use the advantages of the fast Fourier transform (FFT) the integrands can be multiplied by an exponential term, see, for example, [2.66], [2.82] and [2.116]. If the whole integrand is viewed as the product of this exponential term, the corresponding cylindrical function and one further factor, this factor can be expanded with the help of the FFT in a Fourier series. The following integration, carried out term by term, can be performed analytically.

In this book, for the solution of the integral equation (2.18) the field of a current element has to be determined for observation points in the plane $\mid z \mid \ll \lambda$; typical microstrip antennas as well as antennas in media consisting of many layers will be analysed. This is why, in the following, a method will be developed which is not restricted to a special structure. It is based on an integration along the real k_ρ-axis, which avoids the complicated determination of cylindrical functions with complex arguments, as well as on a splitting up of the integrands in order to avoid problems with poor convergence. The development of the following considerations starts with the eqns. (A3.47) to (A3.58) in the range $z < 0$; here it is obviously useful to substitute the Hankel function by the Bessel function. As a preparation for the discussion of the numerical method of solution, the asymptotic behaviour of the integrands will be analysed in the next section.

2.4.3.2 Asymptotic behaviour of the integrands

Inserting

$$|k\rho| \gg |k_i| \tag{2.106}$$

into (2.36) yields

$$k_{zi} \approx -j\,k_\rho \quad . \tag{2.107}$$

The behaviour of the screen coefficients $S_-^{TM/TE}$, the reflection coefficients $\Gamma_i^{TM/TE}$ and of the transmission coefficients $T_i^{TM/TE}$ for $k_\rho \to \infty$ can be deduced by inspection from the eqns. (A2.12), (A2.13), (A3.27), (A3.42) and (A3.43):

$$\lim_{k_\rho \to \infty} S_-^{TM/TE} = 1 \quad , \tag{2.108}$$

$$\lim_{k_\rho \to \infty} \Gamma_i^{TM/TE} = \text{const}(k_\rho) \quad , \tag{2.109}$$

$$\lim_{k_\rho \to \infty} T_i^{TM/TE} = \text{const}(k_\rho) \quad . \tag{2.110}$$

Thus, convergence of the integrals is mainly forced by the exponential terms

$$e^{-j\,k_{zi}\zeta_{i1,2}} \tag{2.111}$$

with

$$\left.\begin{array}{l} \zeta_{i1} = -z, \quad z < 0 \\ \zeta_{i2} = z + 2\,d_i \end{array}\right\} \quad \zeta_{i1/2} > 0 \qquad (2.112)$$

The lesser convergent term determines the total convergence. Since for all regions i it is valid that

$$|z| < d_i \qquad (2.113)$$

it must hold that

$$|z| < |z + 2\,d_i| \quad . \qquad (2.114)$$

Thus, for convergence the term $e^{+j\,k_{zi}z}$ ($z < 0$) is crucial.

2.4.3.3 Improving the convergence

From the explanations in the preceding section it can be seen that the integrals for observation points in the planes $z = 0$ and $|\,z+2d_i\,| = 0$ either do not converge at all or only very poorly. The integrals, however, do exist, as can be seen from the integral presentation of the fields of a Hertzian dipole in a homogeneous space, which for $z = 0$, contains integrals, among others, of the type

$$\int\limits_0^\infty k_\rho^2 J_1(k_\rho \rho)\, dk_\rho \quad . \qquad (2.115)$$

It is true that this integral is divergent; nevertheless, the limit of the integral

$$\lim_{z \to 0} \int\limits_0^\infty k_{z1} k_\rho e^{-j\,k_{z1}z} J_1(k_\rho \rho)\, dk_\rho \qquad (2.116)$$

does exist. Yet, an integral of the type stated above is not suitable for numerical analysis. Unreliable results are also obtained if, for example, eqn. (2.116) is numerically integrated for $z \neq 0$, $|\,z\,| \ll \lambda$. In the following it will be shown that this problem can be solved by splitting up adequate terms. This is done by choosing the exponential term which causes the convergence of the parts evaluated numerically, on the one hand to be as large as possible, and on the other hand so that there is a chance to find analytical solutions for the remaining terms. Which method in each case is preferred depends on the structure to be analysed as well as on the location of the observation point. In the following, for the observation point $z \neq 0$, $|\,z\,| \ll \lambda_0$ is assumed, and a structure as in Fig. 2.7 is chosen. The discussion will be based on the general formula of eqns. (A3.62) and (A3.63), where the Bessel function has to be inserted for Z.

First of all the part $R_{\Psi 2}$ is separated, which corresponds to the field of the structure without a screen ($d_s \to \infty$, $z < 0$), i.e. S_- can be replaced by $S_- = 1$ (the indices TM/TE etc. will largely be dropped in the following):

$$R_{\Psi 2} = 2AM \int\limits_0^\infty f\left(s_{\Psi\xi} e^{+jk_{z0}z} - / + \Gamma_0^{TM/TE} e^{-jk_{z0}(z + 2d_0)}\right) Z\,F\,dk_\rho \quad .$$

$$(2.117)$$

The remaining integrals are

$$R_{\Psi 1} = 2AM \int\limits_0^\infty f\,[S_- - 1]\,\{I\}\,Z\,F\,dk_\rho \quad , \qquad (2.118)$$

with $\{I\}$ representing the term in brackets in eqn. (2.117). From eqn. (A3.27) one obtains for the term $[S_- - 1]$

$$S_- - 1 = -e^{-2jk_{z0}d_s}\,\frac{1 - / + \Gamma_0 e^{-2jk_{z0}d_0}}{1 - / + \Gamma_0 e^{-2jk_{z0}(d_s + d_0)}} \quad . \qquad (2.119)$$

Obviously the integrals of eqn. (2.118) are also convergent for $z = 0$. The rate of convergence is determined by the geometrical data d_s and d_0. For $\Gamma_0^{TM/TE} = 0$, eqn. (2.117) describes the field of a point source in a homogeneous space with the material parameters ε_{r0} and μ_{r0}. The solution of the integral

$$2AM \int\limits_0^\infty f\,s_{\Psi\xi}\,e^{+jk_{z0}z}\,Z\,F\,dk_\rho \qquad (2.120)$$

thus can be given analytically. After subtracting this term from $R_{\Psi 2}$ the integrals of the type

$$R_{\Psi 3} = 2AM \int\limits_0^\infty f\left(-/ + \Gamma_0^{TM/TE} e^{-jk_{z0}(z + 2d_0)}\right) Z\,F\,dk_\rho \qquad (2.121)$$

remain, which for $d_0 \neq 0$ and $z \leq 0$ can be evaluated numerically.

For $d_0 \to 0$, i.e. for the case when the source lies immediately above the interface, further modifications have to be carried out. From eqns. (A2.12) and (A2.13) for the reflection coefficients, with the condition

$$2k_\rho(d_1 - d_0) \gg 1 \quad , \qquad (2.122)$$

it follows that

$$\Gamma_0^{TM}(k_\rho) \approx \frac{\varepsilon(-)_0^1}{\varepsilon(+)_0^1} \quad \text{and} \quad \Gamma_0^{TE}(k_\rho) \approx \frac{\mu(-)_0^1}{\mu(+)_0^1} \qquad (2.123)$$

and with the additional condition $k_\rho \to \infty$

$$\Gamma_0^{TM} \approx \frac{\varepsilon_{r1} - \varepsilon_{r0}}{\varepsilon_{r1} + \varepsilon_{r0}} \quad \text{and} \quad \Gamma_0^{TE} \approx \frac{\mu_{r1} - \mu_{r0}}{\mu_{r1} + \mu_{r0}} \quad . \qquad (2.124)$$

If then

$$Q_1^{TM} + Q_2^{TM} + Q_3 = \qquad (2.125)$$
$$\left(-\Gamma_0^{TM} + \frac{\varepsilon(-)_0^1}{\varepsilon(+)_0^1} \right) + \left(-\frac{\varepsilon(-)_0^1}{\varepsilon(+)_0^1} + \frac{\varepsilon_{r1} - \varepsilon_{r0}}{\varepsilon_{r1} + \varepsilon_{r0}} \right) + \left(-\frac{\varepsilon_{r1} - \varepsilon_{r0}}{\varepsilon_{r1} + \varepsilon_{r0}} \right)$$

and

$$Q_1^{TE} + Q_2^{TE} + Q_3 = \qquad (2.126)$$
$$\left(\Gamma_0^{TE} - \frac{\mu(-)_0^1}{\mu(+)_0^1} \right) + \left(\frac{\mu(-)_0^1}{\mu(+)_0^1} + \frac{\varepsilon_{r1} - \varepsilon_{r0}}{\varepsilon_{r1} + \varepsilon_{r0}} \right) + \left(-\frac{\varepsilon_{r1} - \varepsilon_{r0}}{\varepsilon_{r1} + \varepsilon_{r0}} \right)$$

one obtains

$$Q_1^{TM} = -\frac{\left[\varepsilon(+)_0^1 - \frac{\varepsilon^2(-)_0^1}{\varepsilon(+)_0^1} \right] \Gamma_1^{TM}(k_\rho)}{\varepsilon(+)_0^1 + \varepsilon(-)_0^1 e^{-j2k_{z1}(d_1 - d_0)} \Gamma_1^{TM}(k_\rho)} e^{-j2k_{z1}(d_1 - d_0)}, \qquad (2.127)$$

$$Q_2^{TM} = \frac{2\varepsilon_{r1}\varepsilon_{r0}(\varepsilon_{r1}\mu_{r1} - \varepsilon_{r0}\mu_{r0})}{(\varepsilon_{r0} + \varepsilon_{r1})} \cdot \frac{k^2}{(\varepsilon_{r1}k_{z0} + \varepsilon_{r0}k_{z1})(k_{z0} + k_{z1})}, \qquad (2.128)$$

$$Q_1^{TE} = \frac{\left[\mu(+)_0^1 - \frac{\mu^2(-)_0^1}{\mu(+)_0^1} \right] \Gamma_1^{TE}(k_\rho)}{\mu(+)_0^1 + \mu(-)_0^1 e^{-j2k_{z1}(d_1 - d_0)} \Gamma_1^{TE}(k_\rho)} e^{-j2k_{z1}(d_1 - d_0)}, \qquad (2.129)$$

$$Q_2^{TE} = \frac{2}{\varepsilon_{r1} + \varepsilon_{r0}} \cdot \frac{\mu_{r1}\varepsilon_{r1}k_{z0} - \mu_{r0}\varepsilon_{r0}k_{z1}}{\mu_{r1}k_{z0} + \mu_{r0}k_{z1}} \quad . \qquad (2.130)$$

The integral

$$R_{\Psi 4/5} = 2AM \int\limits_0^\infty f\, Q_{1/2}^{TM/TE}\, e^{-jk_{z0}(z + 2d_0)}\, Z\, F\, dk_\rho \qquad (2.131)$$

can be evaluated numerically also for $d_0 \ll \lambda_0$. The integral

$$R_{\Psi 6} = 2AM \int\limits_0^\infty f \, Q_3 \, e^{-jk_{z0}(z+2d_0)} \, Z \, F \, dk_\rho \qquad (2.132)$$

describes the field of a dipole in a homogeneous space with ε_{r0} and is thus known. As the integral $R_{\Psi 5}$ poorly converges, a further modification is worthwhile: Eqns. (2.128) and (2.130) become for $k_\rho \gg 1$

$$Q_{2\infty}^{TM} \approx -\frac{\varepsilon_{r1}\varepsilon_{r0}(\varepsilon_{r1}\mu_{r1} - \varepsilon_{r0}\mu_{r0}) \, k^2}{(\varepsilon_{r1} + \varepsilon_{r0})^2 \, k_\rho^2} \qquad (2.133)$$

and

$$Q_{2\infty}^{TE} \approx \frac{2}{\varepsilon_{r1} + \varepsilon_{r0}} \frac{\mu_{r1}\varepsilon_{r1} - \mu_{r0}\varepsilon_{r0}}{\mu_{r1} + \mu_{r0}} \quad . \qquad (2.134)$$

If these approximations are put into eqn. (2.131) together with $k_{z0} \approx -jk_\rho$ then, using the abbreviations a, b and x, the following known integrals are obtained:

$$\int\limits_0^\infty J_1(bx) \, e^{-ax} \, dx = \frac{1}{b}\left(1 - \frac{a}{\sqrt{a^2 + b^2}}\right) \quad , \qquad (2.135)$$

$$\int\limits_0^\infty \frac{1}{x} J_1(bx) \, e^{-ax} \, dx = \frac{\sqrt{a^2 + b^2} - a}{b} \quad . \qquad (2.136)$$

This is why, instead of the integral (2.131), first the difference

$$R_{\Psi 7} = 2AM \int\limits_0^\infty f \, (Q_2^{TM/TE} - Q_{2\infty}^{TM/TE}) \, e^{-jk_{z0}(z+2d_0)} \, Z \, F \, dk_\rho \qquad (2.137)$$

is evaluated numerically and then the part which is solved analytically is added.

Thus, for the calculation of the electric field components the integrals (2.118), (2.131) and (2.137) have to be calculated numerically.

The convergence of the integrals (2.118) and (2.131) is determined by the respective exponential terms and thus can be estimated; for the convergence of (2.137), however, no general statements can be given. Many calculated examples, however, show that the numerical integration can be carried out without any problems.

2.4.3.4 Performance of the numerical integration

It was shown in section 2.4.1 that the integrands have branch points at $k_{\hat{m}}$ and k_n. In the case of loss-free layers $\hat{m}$ and n these branch points lie on

the real k_ρ-axis and thus on the path
of integration. Before performing the
numerical integration one has to check
the properties of the integrands in the
vicinity of the branch points. To avoid
this often very complicated procedure
in the case of a multi-layered structure,
the real dielectric constant $\varepsilon_{r\hat{m},n}$ can
be supplemented by a small imaginary
part $\varepsilon''_{r\hat{m},n} = 0.001$. By this the branch
points are moved away from the real
k_ρ-axis to the fourth quadrant.

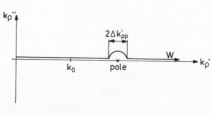

Fig. 2.11   Path of integration for the
numerical integration

As is discussed in detail in section 3.2.4.1, a finite number of poles is located
on the real k_ρ-axis in the case of the microstrip antenna with loss-free dielectric
substrate. Then the exact location of these poles, called $k'_{\rho p}$, is determined.
When performing a numerical integration the range $2\Delta k'_{\rho p}$ is omitted (Fig. 2.11).
The contribution of this range, which can be obtained with the residue theorem,
is added to the result of the numerical integration. The numerical integration
itself is carried out with the help of Simpson's rule. Favourable values for the
increments as well as for the distances $\Delta k'_{\rho p}$ from the poles were determined
through an extensive series of tests.

2.4.4 Approximate equations for very small distances between source and observation point

2.4.4.1 Deviation of simple approximate equations

In the following it is shown that the solutions of the Sommerfeld integrals for
small distances between source and observation points mainly depend on the
properties of the integrands for $k'_\rho \gg k$. In the next section more exact equations are derived. One starts again from the general presentation of the field
components, eqn. (A3.62) with $i = 0$, thus for the layer "0" and, for simplification, for $z < 0$. Again the real k_ρ-axis is chosen as the path of integration, so
that in the following $k''_\rho = 0$ holds. According to eqns. (2.108) and (2.109) the
functions S_- and Γ_0 have limits for $k_\rho \to \infty$. These limits are shown with the
subscript ∞. If one further substitutes

$$f_\infty(k_\rho) = f(k_\rho)\,|_{k_\rho \gg k} \tag{2.138}$$

$$\{I\}_\infty = \{I\}\,|_{k_\rho \gg k}\,, \tag{2.139}$$

then the complete integration can be split up into two intervals:

$$\int\limits_0^\infty f\,S_-\,I\,Z\,dk_\rho = \int\limits_0^\infty (f\,S_-\,I)_\infty\,Z\,dk_\rho + \int\limits_0^{k_{\rho\epsilon}} f\,S_-\,I\,Z\,dk_\rho$$

$$- \int\limits_0^{k_{\rho\epsilon}} (f\,S_-\,I)_\infty\,Z\,dk_\rho + \int\limits_{k_{\rho\epsilon}}^\infty ((f\,S_-\,I)-(f\,S_-\,I)_\infty)\,Z\,dk_\rho\,.$$

$$(2.140)$$

Because S_- and Γ_0 have asymptotes, a number $k_{\rho\epsilon}$ can be found for which the last integral in eqn. (2.140) can be neglected with a given maximum error. With

$$\rho \ll \lambda \text{ and } z \ll \lambda \qquad (2.141)$$

for observation points very close to the source

$$\rho\,k_{\rho\epsilon} \ll 1 \qquad (2.142)$$

is valid. The integrals of the type of the first integral on the right-hand side of eqn. (2.140) can be solved analytically. To this purpose, using eqn. (2.107) one inserts f_∞ according to eqn. (A3.70), $S_{-\infty}$ according to eqn. (2.108), $\Gamma_{0\infty}^{TM}$ and $\Gamma_{0\infty}^{TE}$ according to eqn. (2.124) into the equations for the field components. One gets

$$dE_{\rho0\infty}^{TM} = \frac{2A_E}{k\varepsilon_{r0}} \int\limits_0^\infty j\,k_\rho^2\,\{I\}_\infty^{TM}\,J_1'(k_\rho\rho)\cos\varphi\,dk_\rho \quad , \qquad (2.143)$$

$$dE_{\varphi0\infty}^{TM} = \frac{2A_E}{k\varepsilon_{r0}} \int\limits_0^\infty -j\frac{k_\rho}{\rho}\,\{I\}_\infty^{TM}\,J_1(k_\rho\rho)\sin\varphi\,dk_\rho \quad , \qquad (2.144)$$

$$dE_{z0\infty}^{TM} = \frac{2A_E}{k\varepsilon_{r0}} \int\limits_0^\infty -j\,k_\rho^2\,\{I\}_\infty^{TM}\,J_1(k_\rho\rho)\cos\varphi\,dk_\rho \quad , \qquad (2.145)$$

$$dE_{\rho0\infty}^{TE} = 2A_E k\mu_{r0} \int\limits_0^\infty \frac{1}{j\rho k_\rho}\,\{I\}_\infty^{TE}\,J_1(k_\rho\rho)\cos\varphi\,dk_\rho \quad , \qquad (2.146)$$

$$dE_{\varphi0\infty}^{TE} = 2A_E k\mu_{r0} \int\limits_0^\infty j\,\{I\}_\infty^{TE}\,J_1'(k_\rho\rho)\sin\varphi\,dk_\rho \qquad (2.147)$$

with

$$\{I\}_\infty^{TM} = \left(s_{E\xi}\,e^{k_\rho z} - \Gamma_{0\infty}^{TM}\,e^{-k_\rho(z+2d_0)} \right) \qquad (2.148)$$

and

$$\{I\}_\infty^{TE} = \left(e^{k_\rho z} + \Gamma_{0\infty}^{TE} e^{-k_\rho(z + 2d_0)}\right) \quad , \quad z < 0 \quad . \tag{2.149}$$

Thereby, according to eqn. (A3.72), for the ρ - and φ - components $s_{E\xi}$ has to be set equal to $(+1)$, and for the z-component equal to (-1). We further use the relation

$$J_1'(k_\rho \rho) = \frac{1}{k_\rho} \frac{\partial}{\partial \rho} J_1(k_\rho \rho) \quad , \tag{2.150}$$

the definite integrals

$$\int_0^\infty \frac{1}{x} J_1(bx) e^{-ax} \, dx = \frac{\sqrt{a^2 + b^2} - a}{b} \quad , \tag{2.151}$$

$$\int_0^\infty x J_1(bx) e^{-ax} \, dx = \frac{b}{(a^2 + b^2)^{3/2}} \quad , \tag{2.152}$$

$$\int_0^\infty x^2 J_1(bx) e^{-ax} \, dx = \frac{3ab}{(a^2 + b^2)^{5/2}} \tag{2.153}$$

and eqn. (2.124). Thus, one obtains

$$dE_{\rho 0\infty}^{TM} = 2A_E \frac{1}{k\varepsilon_{r0}} j \frac{\partial}{\partial \rho} \left(\frac{\rho}{r^3} - \frac{\varepsilon_{r1} - \varepsilon_{r0}}{\varepsilon_{r1} + \varepsilon_{r0}} \frac{\rho}{r'^3}\right) \cos\varphi \quad , \tag{2.154}$$

$$dE_{\varphi 0\infty}^{TM} = -2j A_E \frac{1}{k\varepsilon_{r0}} \left(\frac{1}{r^3} - \frac{\varepsilon_{r1} - \varepsilon_{r0}}{\varepsilon_{r1} + \varepsilon_{r0}} \frac{1}{r'^3}\right) \sin\varphi \quad , \tag{2.155}$$

$$dE_{z0\infty}^{TM} = -2j A_E \frac{1}{k\varepsilon_{r0}} \left(\frac{3z\rho}{r^5} - \frac{\varepsilon_{r1} - \varepsilon_{r0}}{\varepsilon_{r1} + \varepsilon_{r0}} \frac{3z'\rho}{r'^5}\right) \cos\varphi \quad , \tag{2.156}$$

$$dE_{\rho 0\infty}^{TE} = -2j A_E k\mu_{r0} \left(\frac{1}{r - z} + \frac{\mu_{r1} - \mu_{r0}}{\mu_{r1} + \mu_{r0}} \frac{1}{r' + z'}\right) \cos\varphi \quad , \tag{2.157}$$

$$dE_{\varphi 0\infty}^{TE} = 2j A_E k\mu_{r0} \frac{\partial}{\partial \rho} \left(\frac{\rho}{r - z} + \frac{\mu_{r1} - \mu_{r0}}{\mu_{r1} + \mu_{r0}} \frac{\rho}{r' + z'}\right) \sin\varphi \tag{2.158}$$

with

$$r = \sqrt{\rho^2 + z^2} \, , \quad r' = \sqrt{\rho^2 + z'^2} \, , \quad z' = z + 2d_0 \quad \text{and} \quad z < 0. \tag{2.159}$$

Note that the TM-parts are proportional to r^{-3} or $(r')^{-3}$ respectively, and that the TE-parts are proportional to $\frac{1}{r}$ or $\frac{1}{r'}$ respectively. Obviously in the nearfield of the source the TE-field can be neglected compared with the TM-field.

The approximations

$$J_1(x) \approx \frac{x}{2} \left.\vphantom{\begin{matrix}a\\b\end{matrix}}\right\} \qquad x \ll 1 \qquad (2.160)$$
$$J_1'(x) \approx \frac{1}{2}$$

are used for the Bessel functions in order to estimate the contributions according to the third term on the right side of eqn. (2.140); for simplicity $z = 0$ and $d_0 = 0$ are chosen. Thus, eqns. (2.143) to (2.147) yield the terms

$$dE_{\rho 0 \infty \varepsilon}^{TM} = A_E \frac{j}{k \varepsilon_{r0}} \frac{k_{\rho \varepsilon}^3}{3} \frac{2\varepsilon_{r0}}{\varepsilon_{r1} - \varepsilon_{r0}} \cos\varphi \quad , \qquad (2.161)$$

$$dE_{\varphi 0 \infty \varepsilon}^{TM} = A_E \frac{-j}{k \varepsilon_{r0}} \frac{k_{\rho \varepsilon}^3}{3} \frac{2\varepsilon_{r0}}{\varepsilon_{r1} + \varepsilon_{r0}} \sin\varphi \quad , \qquad (2.162)$$

$$dE_{z 0 \infty \varepsilon}^{TM} = A_E \frac{j}{k \varepsilon_{r0}} \frac{k_{\rho \varepsilon}^4 \rho}{4} \frac{2\varepsilon_{r1}}{\varepsilon_{r1} + \varepsilon_{r0}} \cos\varphi \quad , \qquad (2.163)$$

$$dE_{\rho 0 \infty \varepsilon}^{TE} = A_E \, k \mu_{r0} \, (-j) \, k_{\rho \varepsilon} \frac{2\mu_{r1}}{\mu_{r1} + \mu_{r0}} \cos\varphi \quad , \qquad (2.164)$$

$$dE_{\varphi 0 \infty \varepsilon}^{TE} = A_E \, k \mu_{r0} \, j \, k_{\rho \varepsilon} \frac{2\mu_{r1}}{\mu_{r1} + \mu_{r0}} \sin\varphi \quad . \qquad (2.165)$$

With $\rho k_{\rho \varepsilon} \ll 1$ these terms can be neglected compared with those given in eqns. (2.154) to (2.156).

For all structures discussed in this study it is valid that the integrands according to the second term on the right side in eqn. (2.140) either are limited or have a finite number of poles and up to two branch points in the range $0 \leq k_\rho' < k_{\rho \varepsilon}$. If it is provided that the branch points are away from the real k_ρ-axis because of small losses, one can give an estimate of these integrals with the help of the same method, as described above. Then it can be seen that they also can be neglected for small values of ρ.

The result is that the nearfield very close to the source can be described approximately by eqns. (2.154) to (2.156). Equations (2.154) to (2.158) have a very simple interpretation if one compares them with the equations for the electric field of a point source in free space: the field consists of a part radiated by the source itself, and a second part radiated by an image source with the amplitude

$$\frac{\varepsilon_{r1} - \varepsilon_{r0}}{\varepsilon_{r1} + \varepsilon_{r0}} \quad . \qquad (2.166)$$

The image is located at $z = -d_0$ (Fig. 2.12). If the point source is located in the interface, then $d_0 = 0$ and thus $r = r'$. Thus, the image source is at the same place as the source itself, and the electric field is determined from the arithmetic average of ε_{r0} and ε_{r1}.

Equations for the fields of sources above the
earth based on image sources can also be found
in papers dealing with the induction method in
geophysical investigations (for example [2.65]).
These differ from those discussed here in so far
that they assume a highly conducting earth for
the derivation, thus

$$|k_1| \gg k \quad . \tag{2.167}$$

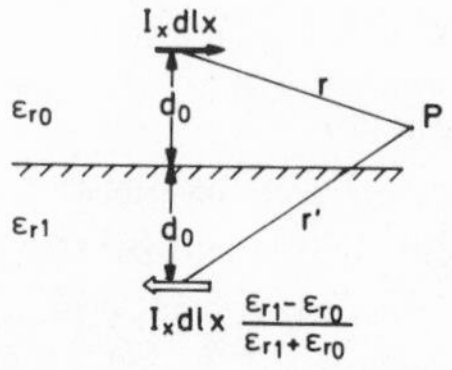

Fig. 2.12 Calculation of the nearfield
with the help of one image
source

In this case the application of eqn. (2.107)
is questionable. Useful approximate equations
can then be derived with the help of "complex" distances between the images
and the original source (see, for example, [2.67], [2.68], [2.71] and [2.118]).

2.4.4.2 Improvement of the approximate equations

Obviously the eqns. (2.154) to (2.156) are only very rough approximations for
the electric field strength because they are independent of the thicknesses of the
layers, and only ε_{r0} and ε_{r1} are involved. Starting from the results of the last
section more exact equations will be derived now by expanding the functions
S_{-}^{TM}, Γ^{TM} and T^{TM} for $k_\rho \gg k$ (k_ρ is real again) and putting parts together
in a suitable way. As in the last section only the range $z < 0$ is considered, for
simplicity. As layers far away from the source have only little influence on the
field (Fig. 2.7), $n = 3$ is chosen. This means that the region $z < -d_2$ is assumed
to be homogeneous. As a first step the approximation valid for k_{z1} in the case
$k_\rho \gg k$ is inserted into eqn. (A2.12). Then one gets

$$\Gamma_0^{TM}(k_\rho) \approx \frac{\varepsilon_{r1} - \varepsilon_{r0}}{\varepsilon_{r1} + \varepsilon_{r0}} \; \frac{1 + \dfrac{\varepsilon_{r2} - \varepsilon_{r1}}{\varepsilon_{r2} + \varepsilon_{r1}} \dfrac{\varepsilon_{r3} - \varepsilon_{r2}}{\varepsilon_{r3} + \varepsilon_{r2}} e^{-2k_\rho(d_2 - d_1)}}{\begin{aligned}1 &+ \dfrac{\varepsilon_{r2} - \varepsilon_{r1}}{\varepsilon_{r2} + \varepsilon_{r1}} \dfrac{\varepsilon_{r3} - \varepsilon_{r2}}{\varepsilon_{r3} + \varepsilon_{r2}} e^{-2k_\rho(d_2 - d_1)} \\ &+ \dfrac{\varepsilon_{r1} + \varepsilon_{r0}}{\varepsilon_{r1} - \varepsilon_{r0}} \dfrac{\varepsilon_{r2} - \varepsilon_{r1}}{\varepsilon_{r2} + \varepsilon_{r1}} e^{-2k_\rho(d_1 - d_0)} \\ &+ \dfrac{\varepsilon_{r1} - \varepsilon_{r0}}{\varepsilon_{r1} + \varepsilon_{r0}} \dfrac{\varepsilon_{r2} - \varepsilon_{r1}}{\varepsilon_{r2} + \varepsilon_{r1}} e^{-2k_\rho(d_1 - d_0)} \\ &+ \dfrac{\varepsilon_{r1} + \varepsilon_{r0}}{\varepsilon_{r1} - \varepsilon_{r0}} \dfrac{\varepsilon_{r3} - \varepsilon_{r2}}{\varepsilon_{r3} + \varepsilon_{r2}} e^{-2k_\rho(d_2 - d_0)} \\ &+ \dfrac{\varepsilon_{r1} - \varepsilon_{r0}}{\varepsilon_{r1} + \varepsilon_{r0}} \dfrac{\varepsilon_{r3} - \varepsilon_{r2}}{\varepsilon_{r3} + \varepsilon_{r2}} e^{-2k_\rho(d_2 - d_0)} \end{aligned}} \cdot \tag{2.168}$$

If the magnitude of the arguments of the exponential functions are much greater
than 1, the denominator can be expanded in a power series. By multiplying the

fractions and rearranging them according to the exponential terms one gets the following series representation:

$$\Gamma_0^{TM}(k_\rho) \approx \frac{\varepsilon_{r1} - \varepsilon_{r0}}{\varepsilon_{r1} + \varepsilon_{r0}} + \frac{\varepsilon_{r2} - \varepsilon_{r1}}{\varepsilon_{r2} + \varepsilon_{r1}} \frac{4\varepsilon_{r1}\varepsilon_{r0}}{(\varepsilon_{r1} + \varepsilon_{r0})^2} e^{-2k_\rho(d_1 - d_0)}$$
$$+ [\quad] e^{-2k_\rho(d_2 - d_0)} + \cdots$$

(2.169)

With the assumption that $2k_\rho(d_s + d_0) \gg 1$, the shielding coefficient (eqn. (A3.27)) can be expanded in a series, too:

$$S_-^{TM}(k_\rho) \approx [1 - e^{-2k_\rho d_s}][1 + \Gamma_0^{TM}(k_\rho) e^{-2k_\rho(d_s + d_0)} + \cdots] \quad . \quad (2.170)$$

Substituting the result from eqn. (2.124), one gets

$$S_-^{TM}(k_\rho) \approx 1 - e^{-2k_\rho d_s} + \frac{\varepsilon_{r1} - \varepsilon_{r0}}{\varepsilon_{r1} + \varepsilon_{r0}} e^{-2k_\rho(d_s + d_0)} + \cdots \quad . \quad (2.171)$$

The following representations for T_1^{TM} and Γ_1^{TM}, which can be determined by similar procedures, are given without derivation:

$$T_1^{TM}(k_\rho) \approx \frac{2\varepsilon_{r0}}{\varepsilon_{r1} + \varepsilon_{r0}} - \frac{\varepsilon_{r2} - \varepsilon_{r1}}{\varepsilon_{r2} + \varepsilon_{r1}} \frac{\varepsilon_{r1} - \varepsilon_{r0}}{\varepsilon_{r1} + \varepsilon_{r0}} \frac{2\varepsilon_{r0}}{\varepsilon_{r1} + \varepsilon_{r0}} e^{-2k_\rho(d_1 - d_0)} + \cdots \quad ,$$

(2.172)

$$\Gamma_1^{TM}(k_\rho) \approx \frac{\varepsilon_{r2} - \varepsilon_{r1}}{\varepsilon_{r2} + \varepsilon_{r1}} + \frac{\varepsilon_{r3} - \varepsilon_{r2}}{\varepsilon_{r3} + \varepsilon_{r2}} \frac{4\varepsilon_{r2}\varepsilon_{r1}}{(\varepsilon_{r2} + \varepsilon_{r1})^2} e^{-2k_\rho(d_2 - d_1)} + \cdots \quad . (2.173)$$

The approximate equations for the nearfield are obtained, considering only the first integral on the right-hand side of eqn. (2.140), as was done in the last section but now substituting for the limits the series representations. Now the integrals which have to be solved correspond to those in eqns. (2.143) to (2.147). Their solutions can be again interpreted with the help of image sources. For the further derivation the ranges $-d_0 < z < 0$ and $-d_1 < z < -d_0$ must be considered separately:

a) $-d_0 < z < 0$

According to definition (A3.46) $T_0^{TM} = 1$ holds; thus, only the multiplication of S_-^{TM} with $\{I\}^{TM}$ has to be carried out. One gets exponential terms in the form of:

$$s_{E\xi-} e_{\nu+} e^{+k_\rho(z - d_{\nu+})} \quad \text{and} \quad e_{\nu-} e^{-k_\rho(z + d_{\nu-})} \quad (2.174)$$

with

$$d_{0+} = 0 \quad ; \quad d_{1+} = 2d_s; \quad d_{2+} = 2(d_s + d_0); \quad \cdots \quad , \quad (2.175)$$
$$d_{0-} = 2d_0 \quad ; \quad d_{1-} = 2d_1; \quad d_{2-} = 2(d_s + d_0); \quad \cdots \quad , \quad (2.176)$$

$$e_{0+} = 1 \; ; \quad e_{1+} = -1; \quad e_{2+} = \frac{\varepsilon_{r1} - \varepsilon_{r0}}{\varepsilon_{r2} + \varepsilon_{r0}}; \quad \cdots \quad , \qquad (2.177)$$

$$e_{0-} = -\frac{\varepsilon_{r1} - \varepsilon_{r0}}{\varepsilon_{r1} + \varepsilon_{r0}}; \quad e_{1-} = -\frac{\varepsilon_{r2} - \varepsilon_{r1}}{\varepsilon_{r2} + \varepsilon_{r1}} \frac{4\varepsilon_{r1}\varepsilon_{r0}}{(\varepsilon_{r1} + \varepsilon_{r0})^2}; \quad e_{2-} = -e_{0-}; \quad \cdots \quad .$$

$$(2.178)$$

Fig. 2.13a displays the structure and Fig. 2.13b the location of the source and of some images. For the derivation of the equations in this section only terms proportional to r^{-3} have been taken into account. Therefore they are valid only for observation points very close to the source. But in contrast to eqns. (2.154) to (2.156) they take into account the influence of the interfaces in the nearfield of the source. Which of the images contribute significantly to the solution is first of all determined by the thicknesses of the layers because these are part of the exponents in eqn. (2.174). For the example shown in Fig. 2.13 it has been assumed that $d_s, d_0, (d_1 - d_0)$ and $(d_2 - d_1)$ have roughly the same value.

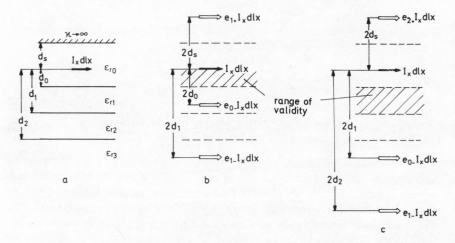

Fig. 2.13 Calculation of the nearfield with the help of image sources

b) $-d_1 < z < -d_0$

In this case the product $S_-^{TM} T_1^{TM} \{I\}^{TM}$ has to be calculated. If one puts the terms together in a suitable way, as in the case a), one obtains the following expressions for $d_{\nu\pm}$ and $e_{\nu\pm}$:

$$d_{0+} = 0; \quad d_{1+} = 2(d_1 - d_0); \quad d_{2+} = 2d_s; \quad \cdots \quad , \qquad (2.179)$$

$$d_{0-} = 2d_1; \quad d_{1-} = 2d_2; \quad d_{2-} = 2(2d_1 - d_0); \quad \cdots \quad , \qquad (2.180)$$

$$e_{0+} = \frac{2\varepsilon_{r0}}{\varepsilon_{r1} + \varepsilon_{r0}}; \quad e_{1+} = -\frac{\varepsilon_{r1} - \varepsilon_{r0}}{\varepsilon_{r1} + \varepsilon_{r0}} \frac{\varepsilon_{r2} - \varepsilon_{r1}}{\varepsilon_{r2} + \varepsilon_{r1}} e_{0+}; \quad e_{2+} = -e_{0+}; \quad \cdots \quad ,$$

$$(2.181)$$

$$e_{0-} = -\frac{\varepsilon_{r2} - \varepsilon_{r1}}{\varepsilon_{r2} + \varepsilon_{r1}} \frac{2\varepsilon_{r0}}{\varepsilon_{r1} + \varepsilon_{r0}} ; \quad e_{1-} = -\frac{\varepsilon_{r3} - \varepsilon_{r2}}{\varepsilon_{r3} + \varepsilon_{r2}} \frac{8\varepsilon_{r2}\varepsilon_{r1}\varepsilon_{r0}}{(\varepsilon_{r2} + \varepsilon_{r1})(\varepsilon_{r1} + \varepsilon_{r0})} ;$$

$$(2.182)$$

$$e_{2-} = -e_{1+}\frac{\varepsilon_{r2} - \varepsilon_{r1}}{\varepsilon_{r2} + \varepsilon_{r1}} ; \quad \ldots \qquad (2.183)$$

Fig. 2.13c shows the image sources for this case.

2.5 Summary of Chapter 2

Chapter 2 gives the theoretical foundations for the analysis of plane antennas in plane stratified structures. The concept of the tensor Green's function has been shown to be an essential formal aid. From a physical point of view Green's function connects the point source located at the source point r with the electric field induced by the point source. The field of an arbitrary current distribution therefore results from the integration over the product of Green's function and the current distribution. An integral equation for the current on conducting surfaces is then obtained by putting the observation point onto the surface and requiring that the tangential electric field be zero. In order to reduce numerical effort for the solution of the integral equation the same restrictions are introduced as is done within the "Thin-wire-theory".

A procedure for the calculation of Green's function was introduced for the first time by Sommerfeld in 1909, for one interface. The formulation he chose has proved to be so ingenious that nearly all papers published up to today are based on Sommerfeld's procedure, as the detailed discussion of the literature in section 2.3.1 shows. The following section 2.3.2 gives a sketch of Sommerfeld's ansatz. The starting point was the Hertz vector $\vec{\Pi}$, which has one non-vanishing component for a current element vertical to the interface and two non-vanishing components for a current element parallel to the interface. Sommerfeld represented the field in both spaces as a superposition of solutions of the wave equation in cylindrical coordinates. Today this representation is called "Sommerfeld integrals". It makes it possible to account for the boundary conditions in the interface in a simple way. It implies, however, that the two components of the Hertz vector are coupled by the boundary conditions. This disadvantage is avoided by the ansatz chosen in 2.3.3, which splits the field into a $TM-$ and a $TE-$part. A very compact formulation is obtained, if all other components are derived from the $E_z^{TM}-$ and the $H_z^{TE}-$component. The single resulting equations are given explicitly, whereby the rather laborious determination of the amplitudes via the boundary conditions is carried out in the appendix (A2 and A3).

The solution of the Sommerfeld integrals takes up much room (section 2.4). Due to the square root in the separation equation the integrands are not unique. Uniqueness is obtained with the help of physical considerations (the radiation condition must be fulfilled). If the ensuing conditions are obeyed, the existence of the integrals is warranted (section 2.4.1). The saddle-point method, which is described in section 2.4.2, is suitable for observation points with a large distance to the source point. The basis of this procedure is a mapping of the $k_\rho-$plane onto the $w-$plane. This removes one branch cut. Thus one obtains unique integrands for a one-sided infinite structure (i.e. structures with a reflector). For structures which are open on both sides a branch cut remains. For the evaluation of the integrands the path of integration is shifted in such a way that it goes through the saddle point. This is a point where the integrand possesses a very steep maximum. Thus the contribution of the saddle point can be given analytically very easily. In addition, contributions from poles, which have been passed during the shift of the path of integration, and, if it exists, the contribution of a branch cut, have to be accounted for.

The description of the numerical integration schemes starts with an investigation of the asymptotic behaviour of the integrands. It turns out that one does not have convergence for observation points in the plane where the source is located. Convergence can be achieved, however, by splitting off suitable terms from the integrands. These terms should preferably be integrable analytically. Altogether, this procedure allows an improvement of the convergence of the integrals for arbitrary observation points.

The numerical integration is carried out along the real axis. For lossless materials there is a finite number of poles on the real axis, the surroundings of which have to be left out in the numerical integration. The contribution of the poles is then calculated with the help of the residue theorem and added to the solution.

For very small distances of source and observation points the field components become singular proportional to the third power of the distances. It is obvious that for this case a numerical evaluation of the integrals is very problematic. On the other hand, one has to account for the effect of the current on itself, i.e. the self-impedance, in order to solve the integral equation. For that one needs very precise values for the observation points with $r \ll \lambda$. In order to derive suitable equations, it is first of all shown that now the behaviour of the integrands for large values of the integration variable k_ρ is decisive. This makes it possible to replace the integrands — excluding the Bessel functions — by their asymptotic approximations. The resulting integrals can be solved analytically. The result can be interpreted as a superposition of the field of the source itself and the field of an image source, where both sources are now located in a homogeneous space. Better approximate equations for the nearfield are obtained by taking into account higher terms in the asymptotic expansion, not only the leading term. The result can again be interpreted in terms of image sources, where one now also has to consider sources, which are reflected at planes located farther away.

The solution methods presented — saddle–point method, numerical integration and approximation schemes — allow efficient calculation of Green's function for any observation point.

CHAPTER 3

Printed antennas

3.1 Survey of the present state of research

The first work on the properties of printed antennas dates from the years 1953 [3.1] and 1955 [3.2]. It was not until 1960 that a further study of the radiation losses of components in microstrip technique appeared [3.3]. The aim of this paper as well as of further investigations at the end of the sixties and the beginning of the seventies was to develop simple equations to give an estimate of Q of resonators in microstrip technique ([3.3] to [3.7]) and of the radiation losses of open–circuit microstrips [3.8]. Here the radiation due to the unscreened structure was considered to be an unavoidable disadvantage of components in microstrip technique. Actual use of the radiation of microstrip components, i.e. the application as antennas, was made for the first time in the early seventies. The structures described by Byron and Munson ([3.9] and [3.10]) consist of straight strips with a length of several wavelengths on a grounded dielectric substrate and with several feed points. The $\lambda_g/2-$ resonators (λ_g = guide wavelength) which are nowadays most frequently used as radiation elements were first described in 1972 by Howell [3.11], then for example by James and by Wiesbeck ([3.12] and [3.13]). The early use of such radiators in arrays is, for example, illustrated in [3.14] and [3.15]. A good summary of the development until the beginning of 1977 is given by James and Wilson [3.16]. The following years saw a rapid development. One should mention the attempts to use logarithmic-periodic structures ([3.17] to [3.20]) and to increase the bandwidth by a variation of the shape of the patch ([3.21] and [3.22]), by locating a number of parasitic elements close to the fed patch antenna ([3.23] to [3.29]), by use of travelling waves ([3.30] and [3.31]) or by increased thickness of the substrate. A circular polarized field is generated by supplying square or circular disc resonators at two feedpoints with a 90° phase shift (e.g. [3.32]) or using geometries which, when supplied only by one feed line, excite two suitable modes ([3.33] to [3.41]). Three review articles published in January 1981 ([3.42] to [3.44]) confirm that with the help of microstrip technique a large number of problems in the field of antenna technique can be resolved.

The early methods to determine the radiation properties of microstrip antennas are based on the abovementioned investigations of radiation losses of microstrip components. In two papers in the early seventies ([3.5] and [3.6]) the current distribution on a $\lambda_g/2$-resonator is assumed to be equal to the distribution on an open-circuit transmission line. The effect of the substrate is described by image sources, that of the dielectric material between the strips and the substrate by polarization currents. The influence of the rest of the dielectric material is neglected. The radiated power is determined with the additional restriction that the width of the strip and the thickness of the substrate are small compared to the wavelength. Whereas in the two papers only equations for the radiation losses are given explicitly, in [3.45] with a similar method, the radiation field also is determined. The results prove to be of only limited use.

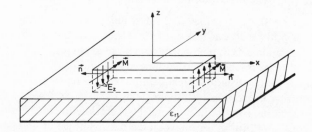

Fig. 3.1 $\lambda_g/2$ Microstrip line and equivalent Huygens sources

The aperture field method has been widely used up to the present day. It is based on an integration over the field in the aperture of the antenna with an appropriately chosen aperture field distribution. The foundation of this method was first discussed by James and Wilson in 1974 [3.12]. By measuring the radiation pattern they found that the farfield of a small $\lambda_g/2$-resonator is similar to that of two magnetic dipoles located at a distance of $\lambda_g/2$ in free space. Therefore they concluded that the two open ends are mainly responsible for the radiation. Starting with the current and field distribution (Fig. 3.1) of a transmission line in resonance, the fields in the faces possess approximately only an E_z-component. Noting that

$$\vec{M} = -\vec{n} \times \vec{E} \tag{3.1}$$

these fields correspond to two magnetic point sources having the same direction. The limitation that the strip has to be very small compared with the wavelength, has been dropped in the next paper [3.46]. Then the radiation field of a $\lambda_g/2$-resonator is equal to that of two magnetic dipoles of finite length. From the radiation field which has thus been determined simple approximate formulas for the directivity and for the elements of an equivalent circuit are developed.

The aperture field method was generalized for resonators of almost any shape by Knoppik in 1976 [3.47]. The method which is used there will be described now for a circular disc antenna. In order to determine the resonance frequency and the field inside the resonator first of all a cavity model is used; its top and bottom are taken as perfectly conductive and its cylinder casing as a perfect magnetic wall (Fig. 3.2, [3.48]). The radiation is determined using the Huygens approach. This consists in determining Huygens sources on the surface of a cylinder encompassing the model. Because of perfect conductivity the condition $\vec{E}_t = 0$ is satisfied on both circular discs. Now the cylinder casing is no longer assumed to be an ideal magnetic wall, but to be a radiating surface with a magnetic surface current

$$\vec{M}_m = -\vec{n} \times \vec{E}_m \quad , \tag{3.2}$$

where $\vec{E}_m$ is the field of the considered mode at the magnetic wall.

The application of this method to a small rectangular $\lambda_g/2$ -resonator yields magnetic surface currents on both broadsides of the resonator (Fig. 3.1) with opposite sign due to the opposite sign of the surface normal vectors. Their effects cancel each other out. Thus the results obtained by James in 1974 [3.12] are confirmed.

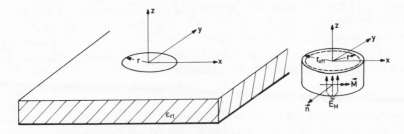

Fig. 3.2 Circular microstrip resonator and corresponding cavity model

The aperture field method is applied in some other papers, for example [3.49] with circular, [3.50] with rectangular and [3.51] and [3.52] with arbitrarily shaped disc resonators. In [3.53] and [3.54] the starting point is the tangential electric field in the surface of the antenna. As this antenna field strongly decreases with increasing distance from the resonator, it is only taken into account within a small region. The application of Huygens' method leads to results which hardly differ from those which have already been mentioned.

The paper of James and Henderson [3.55] can be seen in connection with the aperture field method as well. The radiation field and the amplitudes of the excited surface waves of an open-circuit microstrip line are determined with the mode matching technique, starting with an aperture field corresponding to that of a microstrip line with magnetic walls (Fig. 3.3). In [3.56] the space into which power is radiated is limited by two magnetic walls to the region of $z > 0$, $-w/2 < y < w/2$. This allows the application of the Wiener-Hopf technique. Thus, within this model the radiation field and the surface waves can be determined exactly.

For determination of the input impedance of a narrow $\lambda_g/2$ -resonator Derneryd [3.46] substituted the resonator by a transmission line, which is loaded by an admittance on both ends. Its real part is determined from the radiation losses obtained by integrating the radiation field which has been calculated using the aper-

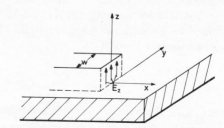

Fig. 3.3   Open - circuited microstrip line

ture field method. The imaginary part of the admittance is obtained from approximate solutions for the fringing capacitance of the circular discs [3.57]. An extension of this method for rectangular and circular disc resonators is presented by Lo et al. [3.58]. Starting from the cavity model with magnetic walls a corrected value for the input impedance is obtained by determining the amplitudes of the modes excited by the feed. This can be done using the mode matching technique [3.59]. Thus, the exact position of the feed point is taken into account. The results obtained in this way correspond very well with those obtained experimentally [3.60]. An application of the cavity model approach to annular sector patches is given in [3.61], to arbitrarily shaped patches in [3.62]. A comparison of the results which were obtained by the various methods using approximate solutions can be found in a paper by Denlinger [3.63]. In [3.64], [3.65] and similarly in [3.66] the cavity model for the circular disc resonator is extended in such a way that the magnetic walls are substituted by walls with finite wall impedance. The imaginary part of the wall impedance is approximately determined by taking into account the fringing capacitance of the circular discs [3.67]. For the real part it is assumed that the radiation is mainly based on the dominant mode.

Although, up to the end of the seventies, the aperture field method and the cavity model have supplied useful results for many cases of technical interest, a more sophisticated approach was needed, as neither method yielded an exact solution of the boundary value problem. In the last 25 years the method of moments has proven to be an effective method for solving electromagnetic boundary value problems. It is particularly simple to apply if one is dealing with a body with a surface of high conductivity, which is surrounded by free space. If the space is partly filled with a dielectric the effect of the dielectric can be described by polarization currents. Both surface currents and polarization currents can be determined using the method of moments. The input impedance of microstrip patch antennas has been calculated in this way by Newman and Tulyathan [3.68]. Here the effect of the grounded dielectric substrate is modelled by doubling the substrate and by image currents. In typical microstrip antennas the distance between the currents in the strips and the image currents which run in the opposite direction is very small. The determination of these currents with the method of moments, therefore, requires a high precision in the numerical analysis. A further disadvantage is that one arrives at very large systems of equations if one takes into account polarization currents which are distributed in space. The application of the method of moments neglecting the polarization currents, which was proposed in 1976 in [3.69], is only a very rough approximation, of course.

An exact solution for the boundary value problem of microstrip structures can formally be given using the so-called spectral domain analysis. Spectral domain analysis was mainly used for the determinaton of propagation constants and of current distributions on screened and unscreened microstrips. The analysis of unscreened rectangular and circular patches was carried out in the beginning of the 1980's [3.70] to [3.80]. Recently rectangular patches have also been considered with dielectric covers [3.81]. The boundary conditions in the interfaces and on the patch are satisfied in the spectral domain applying a two-dimensional Fourier transform. To this end the unknown current distribution on the patch is expanded in "basis functions" which on the one hand are well suited for the problem and on the other hand have a Fourier transform that can be given analytically. Up to now these conditions have limited the application of spectral domain analysis to simple geometries. In these cases, however, the method is very advantageous, as the satisfaction of the boundary conditions in the spectral domain requires the solution of a system of equations instead of an integral equation. In order to determine the coefficients of the system of linear equations one has to deal with integrals which are similar to those appearing in the calculation of Green's function (eqn. (2.62) to (2.73)). Thus, in spectral domain analysis, too, one has to consider the position of branch cuts and poles. Because a shift of a current distribution in the space domain corresponds only to a multiplication by an exponential function in the spectral domain, this method is well suited for the investigation of an array of rectangular patch antennas. The mutual coupling of two elements is calculated in [3.74], [3.77] and in [3.81]; in [3.82] - [3.84] infinite phased arrays and in [3.85] finite phased arrays are analysed. In [3.86] the method is extended to finite arrays with a dielectric cover.

In order to emphasize the difference between spectral domain analysis and the method on which this study is based, we once more briefly refer to the preceding chapter. By calculating Green's function with the help of the integral representation (eqn. 2.18) we satisfy the boundary conditions at the interfaces in the spectral domain. The fitting of the solution to the condition $\vec{E}_{tan} = 0$ on the printed circuit, i.e. the solution of the integral eqn. (2.18) is carried out in the space domain. Once Green's function has been determined for a given layered structure, more complicated setups such as curved dipoles or groups of dipoles of different lengths can be handled without much additional work. To stress the difference between this method and spectral domain analysis the expression "space domain analysis" will be used in the following.

The first work on the radiation properties of printed antennas based on space domain analysis was published at the end of 1979 and at the beginning of 1980 ([3.87] to [3.90]). Instead of the representation by TM- and TE-waves which is used in the present work, some other studies (e.g. [3.87]) use Hertzian vectors. In the following years with the help of space domain analysis many aspects concerning printed antennas were analysed: The current distribution and input impedance ([3.91] and [3.92]), the excitation of surface waves ([3.93] to [3.96]), the effect of thick substrates ([3.97] to [3.99]), the coupling between dipoles ([3.100] to [3.102]), the influence of a dielectric cover ([3.103] to [3.107]), transient problems ([3.108] to [3.109]), arrays of dipoles ([3.110] to [3.114]) and recently the excitation of leaky waves [3.115]. Analytical and numerical details are discussed in [3.116], and an overview of the integral equation technique is given in [3.117]. More references and details can be found in the books "Microstrip antennas" by Bahl and Bhartia [3.118] "Microstrip antenna theory and design" by James, Hall and Wood [3.119], further in the contribution "A dynamical radiation model for microstrip structures" by Mosig and Gardiol [3.120].

In continuation of this work, first of all the general equations for Green's functions, i.e. the field of a point source, will be specialized for the case of an unscreened microstrip structure in the next sections. It is assumed that the antenna, consisting of a printed circuit on a grounded dielectric substrate, is covered with a further dielectric layer (Fig. 3.4). This extension makes it

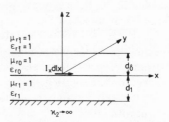

Fig. 3.4 Microstrip antenna with a dielectric cover

possible to analyse the effect which for example a protective coating or layer of ice have on the properties of a microstrip antenna. Note that the equations for such a three layer problem can still be given explicitly and discussed with an amount of work that is reasonable. The printed circuit is excited by one or more voltage sources. The results will then be used to investigate the radiation properties of a dipole of finite length and of an array of dipoles, the elements of which are excited by radiation coupling.

3.2 Field of a point source on a grounded substrate with a cover

3.2.1 Equations for the field components

Starting from the general structure in Fig. 2.1, $\hat{m} = 1$ and $n = 2$ have to be chosen in the case of a microstrip antenna with a dielectric cover. Thus, the source is at first located in the layer "zero", then, with d_0 approaching zero it is in the interface between the substrate and the cover (Fig. 3.4). Above the cover there is free space, hence

$$k_{\hat{1}} = \omega\sqrt{\mu_0\varepsilon_0} = k \qquad (3.3)$$

and

$$k_{z\hat{1}} = \sqrt{k^2 - k_\rho^2} = k_z \quad . \qquad (3.4)$$

The grounding of the substrate is taken into account by $| k_2 | \to \infty$. For further discussions it appears appropriate to replace the exponential functions in the equations for the field components partly by trigonometric functions. In the following, only the equations for the $E_z^{TM}-$ and H_z^{TE}-components in all the three layers are given; the equations for the remaining components can be determined by the eqns. (2.56) to (2.59):

$$dE_{z\hat{1}}^{TM} = \int_{-\infty}^{\infty} B_{\hat{1}} e^{-jk_z z} H_1^{(2)}(k_\rho\rho)\cos\varphi \, dk_\rho \quad , \qquad (3.5)$$

$$dH_{z\hat{1}}^{TE} = \int_{-\infty}^{\infty} -D_{\hat{1}} e^{-jk_z z} H_1^{(2)}(k_\rho\rho)\sin\varphi \, dk_\rho \quad , \qquad (3.6)$$

$$dE_{z0}^{TM} = \int_{-\infty}^{+\infty} \frac{B_{\hat{1}}}{k_{z0}} \left[\frac{k_{z0}}{\varepsilon_{r0}}\cos(k_{z0}(d_{\hat{0}} - z)) + jk_z\sin(k_{z0}(d_{\hat{0}} - z)) \right]$$
$$e^{-jk_z d_{\hat{0}}} H_1^{(2)}(k_\rho\rho)\cos\varphi \, dk_\rho \quad , \qquad (3.7)$$

$$dH_{z0}^{TE} = \int_{-\infty}^{+\infty} -D_{\hat{1}} \left[\cos(k_{z0}(d_{\hat{0}} - z)) + j\frac{k_z}{k_{z0}}\sin(k_{z0}(d_{\hat{0}} - z)) \right]$$
$$e^{-jk_z d_{\hat{0}}} H_1^{(2)}(k_\rho\rho)\sin\varphi \, dk_\rho \quad , \qquad (3.8)$$

$$dE_{z1}^{TM} = \int_{-\infty}^{+\infty} j\frac{B_{\hat{1}}}{k_{z1}}\frac{\cos(k_{z1}(d_1 + z))}{\sin(k_{z1}d_1)} \left[k_z\cos(k_{z0}d_{\hat{0}}) + j\frac{k_{z0}}{\varepsilon_{r0}}\sin(k_{z0}d_{\hat{0}}) \right]$$
$$e^{-jk_z d_{\hat{0}}} H_1^{(2)}(k_\rho\rho)\cos\varphi \, dk_\rho \quad , \qquad (3.9)$$

$$dH_{z1}^{TE} = \int\limits_{-\infty}^{+\infty} -D_{\hat{1}} \frac{\sin(k_{z1}(d_1 + z))}{\sin(k_{z1}d_1)} \left[\cos(k_{z0}d_{\hat{0}}) + j\frac{k_z}{k_{z0}} \sin(k_{z0}d_{\hat{0}}) \right]$$

$$e^{-jk_z d_{\hat{0}}} H_1^{(2)}(k_\rho \rho) \sin\varphi \, dk_\rho \quad . \tag{3.10}$$

This presentation is chosen so as to make the equations as simple as possible in the region above the layers. For this reason the amplitudes are normalized by $B_{\hat{1}}$ or $D_{\hat{1}}$, respectively, the calculation of which can be carried out with the help of the eqns. (A3.62) to (A3.77). One gets

$$B_{\hat{1}} = \frac{I_x dl_x k_\rho^2}{4\pi\omega\varepsilon_0\varepsilon_{r1}} \frac{\sin(k_{z1}d_1) \, e^{jk_z d_{\hat{0}}}}{j\frac{1}{\varepsilon_{r1}} \cos(k_{z0}d_{\hat{0}}) \, \sin(k_{z1}d_1) + j\frac{1}{\varepsilon_{r0}} \frac{k_{z0}}{k_{z1}} \sin(k_{z0}d_{\hat{0}}) \, \cos(k_{z1}d_1)}$$

$$\overline{-\frac{\varepsilon_{r0}}{\varepsilon_{r1}} \frac{k_z}{k_{z0}} \sin(k_{z0}d_{\hat{0}}) \, \sin(k_{z1}d_1) + \frac{k_z}{k_{z1}} \cos(k_{z0}d_{\hat{0}}) \, \cos(k_{z1}d_1)} \, , \tag{3.11}$$

and

$$D_{\hat{1}} = -\frac{I_x dl_x k_\rho^2}{4\pi} \frac{\sin(k_{z1}d_1) \, e^{jk_z d_{\hat{0}}}}{k_{z1} \cos(k_{z1}d_1) \, \cos(k_{z0}d_{\hat{0}}) + j\frac{k_z k_{z1}}{k_{z0}} \cos(k_{z1}d_1) \, \sin(k_{z0}d_{\hat{0}})}$$

$$\overline{-k_{z0} \sin(k_{z0}d_{\hat{0}}) \, \sin(k_{z1}d_1) + jk_z \cos(k_{z0}d_{\hat{0}}) \, \sin(k_{z1}d_1)} \, . \tag{3.12}$$

The equations become considerably easier in the case of a microstrip antenna without a dielectric cover [3.73]:

$$dE_{z0}^{TM} = -j\frac{I_x dl_x}{4\pi\omega\varepsilon_0} \int\limits_{-\infty}^{+\infty} k_\rho^2 \frac{1}{1 - j\frac{k_z}{k_{z1}}\varepsilon_{r1}\cot(k_{z1}d_1)} e^{-jk_z z} H_1^{(2)}(k_\rho\rho) \cos\varphi \, dk_\rho \, ,$$

$$\tag{3.13}$$

$$dH_{z0}^{TE} = -j\frac{I_x dl_x}{4\pi} \int\limits_{-\infty}^{+\infty} \frac{k_\rho^2}{k_z} \frac{1}{1 - j\frac{k_{z1}}{k_z}\cot(k_{z1}d_1)} e^{-jk_z z} H_1^{(2)}(k_\rho\rho) \sin\varphi \, dk_\rho \, ,$$

$$\tag{3.14}$$

$$dE_{z1}^{TM} = j\frac{I_x dl_x}{8\pi\omega\varepsilon_0} \int\limits_{-\infty}^{+\infty} \frac{k_\rho^2 k_z}{k_{z1}} \frac{1 - j\cot(k_{z1}d_1)}{1 - j\frac{k_z}{k_{z1}}\varepsilon_{r1}\cot(k_{z1}d_1)} \tag{3.15}$$

$$\left(e^{jk_{z1}z} + e^{-jk_{z1}(2d_1 + z)} \right) H_1^{(2)}(k_\rho\rho) \cos\varphi \, dk_\rho \, ,$$

$$dH_{z1}^{TE} = -j\frac{I_x dl_x}{8\pi} \int\limits_{-\infty}^{+\infty} \frac{k_\rho^2}{k_z} \frac{1 - j\cot(k_{z1}d_1)}{1 - j\frac{k_{z1}}{k_z}\cot(k_{z1}d_1)} \tag{3.16}$$

$$\left(e^{jk_{z1}z} - e^{-jk_{z1}(2d_1 + z)} \right) H_1^{(2)}(k_\rho\rho) \sin\varphi \, dk_\rho \quad .$$

3.2.2 Analytical properties of the integrands

In this passage a few analytical properties of the integrands will be discussed in continuation of the ideas put forward in section 2.4.1 and 2.4.3. To start with, it is assumed that all dielectrics are lossfree, i.e. all k_i are real. Then it is straightforward to see that two branch points ($k_{\rho b1} = \pm k$) lie on the real k_ρ-axis and, if the Hankel function is used, even one more ($k_{\rho b3} = 0$) lies on the real axis. To determine the poles, the zeros of the denominators of B_i and D_i have to be calculated. In the case of a typical microstrip antenna the functions are simple; graphical solutions are described in the literature ([A1.11], p. 112). In order to facilitate the discussion of the analytical properties when dealing with structures with several layers it is useful to represent graphically the essential parts of the integrands in the complex k_ρ-plane. For this purpose the branch cuts are drawn on the real k_ρ-axis towards $\pm\infty$ (as in Fig. 2.8) starting from the branch points, and the sign of the square root is chosen in such a way that the condition $Re\{k_z\} > 0$ (eqn. (2.75)) is fulfilled in the entire upper plane. The sign of the imaginary part of k_z can be found by applying the method shown in Fig. 2.8. Both conditions, eqns. (2.75) and (2.76), are valid for the first and the third quadrant. As a result of the general equation for the field components eqn. (A3.62), the location of the poles can be determined by means of discussing the analytical properties of the product

$$P_{STI}^{TM/TE} = S^{TM/TE} \, T_i^{TM/TE} \, \{I\}^{TM/TE} \quad . \tag{3.17}$$

For this it is irrelevant, of course, which numerical value is used for z. For this reason, and to simplify the equation, in all examples shown, $z = 0$ and thus $T_0^{TM/TE} = 1$.

The first example in Fig. 3.5 shows $| P_{STI}^{TM/TE} |$ for a lossfree dielectric sheet with $\varepsilon_{r1} = 10$ (for example ceramic) and $d_1 = 6$ mm on a perfect ground in the range

$$-15.0 \, \tfrac{1}{cm} \; < k_p' < +15.0 \, \tfrac{1}{cm} \quad ,$$

$$-40.0 \, \tfrac{1}{cm} \; < k_p'' < +40.0 \, \tfrac{1}{cm} \quad .$$

The source is situated on the dielectric sheet ($d_0 = 0$), the wavelength is $\lambda = 3$ cm. One recognizes sets of poles in the second and fourth quadrant parallel to the imaginary axis. As a further illustration an additional figure shows an area close to the origin of coordinates on an enlarged scale. One can see that further poles lie on the real axis in the range $k_\rho' > k$. As a more detailed analysis of for example P_{STI}^{TM} shows, two poles have to be assigned to the first quadrant ("above" the branch cut as shown in Fig. 2.8), one pole has to be assigned to the fourth quadrant ("beneath" the branch cut). In Fig. 3.6 $| P_{STI}^{TM/TE} |$ has been calculated for the case of a protective Teflon layer ($\varepsilon_{r0} \approx 2.5$) on the substrate. In the second and fourth quadrant (only the

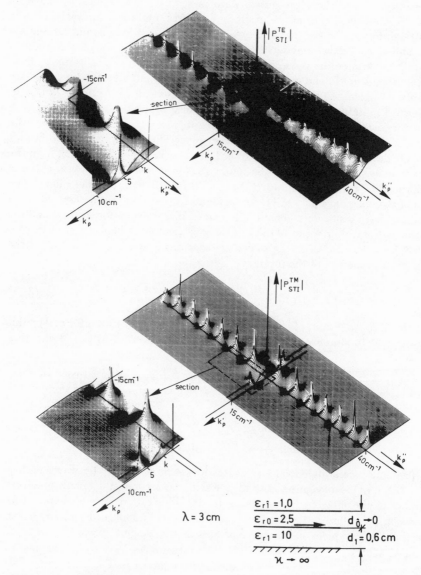

Fig 3 5 Plot of $\left|P_{STI}^{TE/TM}\right|$ in the k_ρ -plane showing the singularity of the integrands

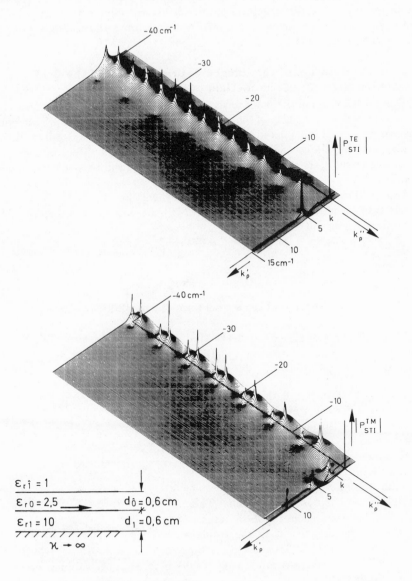

Fig. 3.6 Plot of $\left|P_{STI}^{TE/TM}\right|$ in the k_p -plane showing the singularity of the integrands

fourth being shown) the $P_{STI}^{TM/TE}$ have two sets of poles located close to each other, which run parallel to the imaginary axis, and have poles on the real axis. In all the other lossfree structures studied it was also found that all poles are either situated in the second and fourth quadrant or on the real axis on either side of the branch cut.

3.2.3 Interpretation of the integral representation of the field with the help of the method of steepest descent

The contribution of the poles to the solution of the Sommerfeld integrals can be interpreted with the help of the method of steepest descent [2.89]. To simplify the presentation it is supposed that all poles are far enough away from the saddle point w_s so that the second term of eqn. (2.98) can be used. Based on the above results concerning the positions of the poles two cases will be discussed in more detail:

a) One pole is situated on the positive real k_ρ-axis, i.e. $k_{\rho p} = k'_{\rho p} > 0$. Using the transformation according to eqn. (2.79) the position of this pole in the w-plane comes out as

$$w_p = \tfrac{\pi}{2} + j w''_p \tag{3.18}$$

where

$$k'_{\rho p} = k \cosh w''_p \quad . \tag{3.19}$$

Thus, for the exponent of the second term of eqn. (2.98) one gets

$$e^{-j k r' \cos(w_p - \vartheta)} = e^{-j k r' \sin\vartheta \, \cosh w''_p} \, e^{-k r' \cos\vartheta \, \sinh w''_p}$$
$$= e^{(-j k'_{\rho p} \rho - k z' \sinh w''_p)} \quad . \tag{3.20}$$

For $k'_{\rho p} > 0$, $w''_p > 0$ this equation describes a surface wave propagating parallel to the surface with the propagation velocity $\omega/k'_{\rho p}$, the field components of which decay exponentially in the z-direction. The condition $w''_p > 0$ implies, according to Fig. 2.8, that for the pole $Re\{k_z\} = 0$, $Im\{k_z\} < 0$ holds. Thus, it is fixed that such a pole is situated "above" the branch cut in the k_ρ-plane (Fig. 3.7). For $w''_p < 0$, i.e. for poles "beneath" the branch cut, eqn. (3.20) results in an exponential increase of the amplitude.

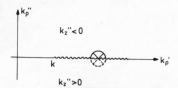

Fig. 3.7   Pole located "above" the branch cut

b) A pole is situated in the fourth quadrant, i.e. it is

$$k_{\rho p} = k'_{\rho p} + j k''_{\rho p}$$

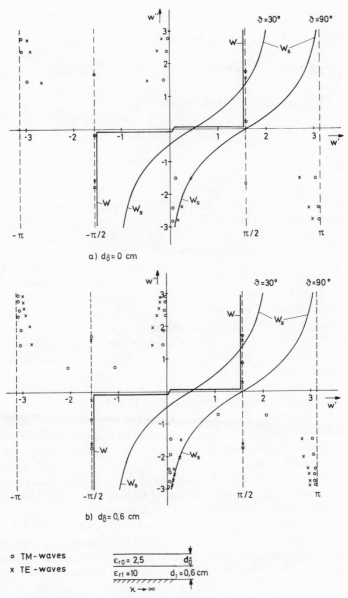

Fig. 3.8 Plot of the poles in the w-plane

with $k'_{\rho p} > 0$, $\quad k''_{\rho p} < 0$ and $Re\{k_z\} > 0$. With the help of Fig. 2.8 one finds in this case that the pole in the w-plane lies in the strip

$$0 < w'_p < \pi/2 \ \text{ and } w''_p < 0 \quad . \tag{3.21}$$

One gets for the exponential function

$$e^{\left(-jk\cosh w''_p \, (z'\cos w'_p \, + \rho\sin w'_p \,) - k\sinh w''_p \, (z\sin w'_p \, - \rho\cos w'_p \,)\right)} \ . \tag{3.22}$$

Under the conditions (3.21) it follows that

$$\sin w'_p \, > 0, \quad \cos w'_p \, > 0 \tag{3.23}$$

and

$$\sinh w''_p \, < 0 \quad . \tag{3.24}$$

Thus, the exponential function describes a plane wave, the direction of propagation of which is given by

$$\vec{n} = \vec{u}_z \cos w'_p \, + \vec{u}_\rho \sin w'_p \quad . \tag{3.25}$$

It decays exponentially in the ρ-direction and increases exponentially in the z-direction. This type of wave is called a leaky wave. It does not satisfy the radiation condition, which was to be expected because of the position of its corresponding pole in the fourth quadrant.

Similarly it is possible to discuss the position of the poles with $k'_{\rho p} < 0$, $k''_{\rho p} = 0$ and the poles with $k'_{\rho p} < 0$, $k''_{\rho p} > 0$ in the w-plane, and how their contribution according to eqn. (2.98) is to be interpreted.

Two examples for the position of the poles in the w-plane are shown in Fig. 3.8a and 3.8b, where the same geometrical and material data have been used as in Fig. 3.5 and 3.6. We have also drawn the original path of integration W, which has been transformed into a path in the w-plane, and the path of the steepest descent W_s passing through the saddle point according to eqn. (2.94) for $\vartheta = 30^o$ and $\vartheta = 90^o$.

According to eqn. (2.98) only those poles have to be taken into account which are encountered during the deformation of W into W_s. These must be situated in the first or fourth quadrant of the k_ρ- plane, and therefore represent surface waves or leaky waves according to the explanations in subsections a) and b). For the latter the compatibility of the exponential increase of the amplitude with z, as follows from eqn. (3.22), with the radiation condition, has to be proved: The term describing the attenuation in eqn. (3.22) can also take the following form:

$$e^{-kr' \sinh w''_p \, sin(w'_p - \vartheta)} \quad . \tag{3.26}$$

According to Fig. 3.8, besides $w_p'' < 0$ also $w_p' < \vartheta$ is satisfied for leaky wave poles situated between W and W_s. Therefore the amplitude decreases exponentially with growing r' at a constant angle ϑ. With a given ρ, the contribution of a leaky wave pole may only be taken into account up to a maximum value z, which can be determined from eqn. (2.96). It further follows from eqn. (3.26) that the attenuation increases exponentially with $\sinh w_p''$. Thus, only poles close to the real w-axis are relevant.

Interpreting the Sommerfeld integrals with the help of the method of steepest descent one has to take into account that $\rho \gg \lambda$ has been assumed in the derivation of this method. Splitting up of the field at a certain point into the contribution given by eqn. (2.98), i.e. the field of a source in a homogeneous space, the reflected field from an interface, the field of surface and leaky waves, therefore is correct only in the sense of an asymptotic solution. For the solution of the integral eqn. (2.18), the shape of the field in the plane $z = 0$ and for $\rho \lesssim 2\lambda$ is of relevance. In section 3.2.5 we will therefore present some results for the amplitude and the phase of the electric field in this range. They show that surface waves dominate for $\rho \gtrsim \lambda/2$ and then determine the mutual coupling. Therefore, in section 3.2.4.1 and 3.2.4.2 the properties of excited surface waves will be discussed. For the calculation of the radiation pattern in the farfield of an antenna, the distances ρ and z for an observation point (r, ϑ, φ) can always be chosen in such a way that the conditions $z \gg \lambda$ and $\rho \gg \lambda$ are fulfilled. Then both surface and leaky waves are negligible, and we are left with the first term of eqn. (2.98), i.e. the contribution of the saddle point (eqns. (2.100) - (2.105)). As this saddle point describes the power which is radiated into the space, the corresponding field shall be called "space wave". The derivations of the equations for the space wave are carried out in section 3.2.5.

3.2.4 Surface waves

3.2.4.1 Propagation constants of surface waves

According to eqn. (3.20) the propagation constant is determined by the position of the poles on the real k_ρ-axis, $k'_{\rho p}$. In order to determine the poles, the zeros of the denominators of $B_{\hat{1}}$ and $D_{\hat{1}}$ have to be calculated:

$$B_{\hat{1}N} = j\frac{1}{\varepsilon_{r1}}\cos(k_{z0}d_{\hat{0}})\,\sin(k_{z1}d_1) + j\frac{1}{\varepsilon_{r0}}\frac{k_{z0}}{k_{z1}}\sin(k_{z0}d_{\hat{0}})\,\cos(k_{z1}d_1) \qquad (3.27)$$

$$- \frac{\varepsilon_{r0}}{\varepsilon_{r1}}\frac{k_z}{k_{z0}}\sin(k_{z0}d_{\hat{0}})\,\sin(k_{z1}d_1) + \frac{k_z}{k_{z1}}\cos(k_{z0}d_{\hat{0}})\,\cos(k_{z1}d_1) = 0 \quad ,$$

$$D_{\hat{1}N} = k_{z1}\cos(k_{z1}d_1)\,\cos(k_{z0}d_{\hat{0}}) + j\frac{k_z k_{z1}}{k_{z0}}\cos(k_{z1}d_1)\,\sin(k_{z0}d_{\hat{0}}) \qquad (3.28)$$

$$- k_{z0}\sin(k_{z0}d_{\hat{0}})\,\sin(k_{z1}d_1) + jk_z\cos(k_{z0}d_{\hat{0}})\,\sin(k_{z1}d_1) = 0 \quad .$$

A surface wave propagates if $k'_{\rho p} \geq k$. Now the cutoff wavelength $\lambda_{cn}^{TM/TE}$ or the cutoff frequency

$$f_{cn}^{TM/TE} = \frac{1}{\sqrt{\mu_0\varepsilon_0}\,\lambda_{cn}^{TM/TE}} \qquad (3.29)$$

can be determined. With

$$k_z = \sqrt{k^2 - k_\rho^2} = 0 \qquad (3.30)$$

the equations for $\lambda_{cn}^{TM/TE}$ are obtained from eqns. (3.27) and (3.28):

$$\mathrm{tg}\Big(\frac{2\pi}{\lambda_{cn}^{TM}}d_1\sqrt{\varepsilon_{r1}-1}\,\Big) = -\frac{\sqrt{\varepsilon_{r0}-1}}{\sqrt{\varepsilon_{r1}-1}}\frac{\varepsilon_{r1}}{\varepsilon_{r0}}\,\mathrm{tg}\Big(\frac{2\pi}{\lambda_{cn}^{TM}}d_{\hat{0}}\sqrt{\varepsilon_{r0}-1}\,\Big) \quad (TM_n-\text{wave})\,,$$

$$(3.31)$$

$$\mathrm{tg}\Big(\frac{2\pi}{\lambda_{cn}^{TE}}d_1\sqrt{\varepsilon_{r1}-1}\,\Big) = \frac{\sqrt{\varepsilon_{r1}-1}}{\sqrt{\varepsilon_{r0}-1}}\,\mathrm{ctg}\Big(\frac{2\pi}{\lambda_{cn}^{TE}}d_{\hat{0}}\sqrt{\varepsilon_{r0}-1}\,\Big) \quad (TE_n-\text{wave})\,.$$

$$(3.32)$$

From eqn. (3.31) it follows that the TM_1-wave has the cutoff frequency

$$f_{c1}^{TM} = 0 \quad , \qquad (3.33)$$

whereas due to eqn. (3.32) the TE_1-wave has a finite cutoff frequency, $f_{c1}^{TE} \neq 0$. Thus, the TM_1-wave is the dominant mode.

Generally, the solutions of the equations for cutoff frequencies and propagation constants have to be found numerically. Some results will be discussed in the following, choosing low ($\varepsilon_r = 2.5$) and high ($\varepsilon_r = 10$) values for the dielectric constant of the substrate and the cover. In addition, the thicknesses of both layers were varied. Fig. 3.9 shows results for the cutoff wavelength.

The cutoff wavelengths increase in all cases with growing thicknesses of the layers. Until now microstrip antennas have mostly been designed in such a way as to prevent excitation of surface waves, if possible. It can be read off Fig. 3.9 which modes can be excited, for example, after applying a protective layer.

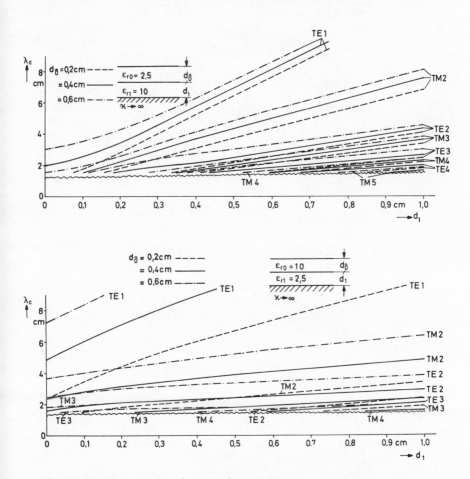

Fig. 3.9 Cutoff wavelengths of some surface waves

Fig. 3.10a and 3.10b show the results for the propagation constants. If $\varepsilon_{r0} < \varepsilon_{r1}$ the curves roughly resemble those for a microstrip antenna without a cover ($d_{\hat{0}} = 0$): In the case of thin layers the propagation constant is not much larger than k; if the thicknesses of the layers are increased it approaches the wave number of the substrate.

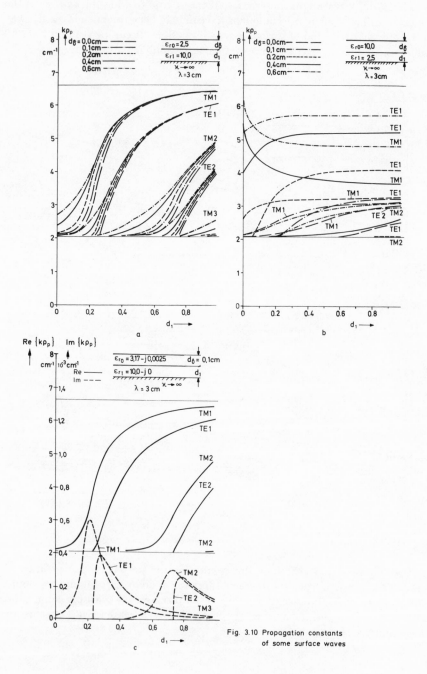

Fig. 3.10 Propagation constants
of some surface waves

If $\varepsilon_{r0} > \varepsilon_{r1}$ the situation is different altogether: Although the curves approach asymptotes for thick substrates, now the position of the asymptotes depends on k_0, k_1 and on the mode under consideration. Especially if $\varepsilon_{r0} \gg \varepsilon_{r1}$ and if the cover is thick, the figure of $k_{\rho p}$ for thin substrates can be larger than for thick ones.

In sections 3.2.2, 3.2.3 and also in this one it has been assumed that the cover consists of lossfree material. This is not true, however, if the surface of the antenna is covered for example with a layer of ice, dust or water. Fig. 3.10c shows, as an example, the influence of an ice layer on the propagation constant of surface waves. Since ice shows only low losses at 10 GHz, the real part of $k_{\rho p}$ is mostly determined by the real part of the dielectric constants. Because of the losses, $k_{\rho p}$ has a negative imaginary part. In the complex k_ρ-plane the pole is situated on the lower sheet if a branch cut as in Fig. 2.8 is chosen, thus $Re\{k_z\} < 0$. Based on the same considerations as in section 3.2.2 one can conclude that the pole is situated in the w- plane in the range $\pi/2 < w' < \pi$, $w'' > 0$ and that therefore its contribution describes the field of an attenuated surface wave, as was to be expected. From Fig. 3.10c one can see that attenuation is small for a mode if its phase velocity is mainly determined either by $\varepsilon_{ri} = 1$ or by ε_{r1}. In between $| k_{\rho p}'' |$ has a maximum. Altogether the results show that even thin ice layers make it necessary to take into account additional surface waves.

In the special case of very thin layers the equations that determine the propagation constants (eqns. (3.27) and (3.28)) can be simplified considerably. This case is of interest as layers have usually been made so thin for microstrip antennas that surface waves are excited as little as possible. With

$$d_i \ll \lambda, \qquad i = 1, \hat{0} \tag{3.34}$$

the approximations $\cos(k_{zi}d_i) \approx 1$ and $\sin(k_{zi}d_i) \approx k_{zi}d_i$ hold. Then one obtains for the TM-waves

$$j\frac{1}{\varepsilon_{r1}}k_{z1}d_1 + j\frac{k_{z0}}{\varepsilon_{r0}k_{z1}}k_{z0}d_{\hat{0}} - \frac{\varepsilon_{r0}k_z}{\varepsilon_{r1}k_{z0}}k_{z0}d_{\hat{0}}k_{z1}d_1 + \frac{k_z}{k_{z1}} = 0 \tag{3.35}$$

and for TE-waves

$$k_{z1}k_{z0} + jk_zk_{z1}k_{z0}d_{\hat{0}} - k_{z0}^3 d_{\hat{0}}k_{z1}d_1 + jk_zk_{z0}k_{z1}d_1 = 0 \quad . \tag{3.36}$$

If k_{z0} and k_{z1} are substituted by k_z eqn. (3.35) is a quadratic equation for k_z. If further

$$k_z^* = jk_z \tag{3.37}$$

is put because of $k_\rho' > k$, this quadratic equation reads

$$k_z^{*2} + \frac{\varepsilon_{r0}\varepsilon_{r1}}{d_1\varepsilon_{r0} + d_{\hat{0}}\varepsilon_{r1}}k_z^* + k^2\left(1 - \frac{\varepsilon_{r0}\varepsilon_{r1}(d_{\hat{0}} + d_1)}{d_1\varepsilon_{r0} + d_{\hat{0}}\varepsilon_{r1}}\right) = 0 \quad . \tag{3.38}$$

Its approximate solution yields

$$k_{\rho p1}^{TM} = k\sqrt{1 + \left(k\left(d_1 + d_{\hat{0}}\right)\left(1 - \frac{\varepsilon_{r1}d_{\hat{0}} + \varepsilon_{r0}d_1}{\varepsilon_{r0}\varepsilon_{r1}(d_{\hat{0}} + d_1)}\right)\right)^2} \quad . \tag{3.39}$$

If the same calculations are carried out for a microstrip antenna without a cover, i.e. with $d_{\hat{0}} = 0$ or $\varepsilon_{r0} = \varepsilon_{ri} = 1$, one gets

$$k_{\rho p1}^{TM} = k\sqrt{1 + \left(kd_1(1 - \frac{1}{\varepsilon_{r1}})\right)^2} \quad . \tag{3.40}$$

The term

$$\varepsilon_{reff} = \frac{\varepsilon_{r0}\varepsilon_{r1}(d_{\hat{0}} + d_1)}{\varepsilon_{r1}d_{\hat{0}} + \varepsilon_{r0}d_1} \tag{3.41}$$

can therefore be considered as an effective ε_r.

Eqn. (3.36) reads after cancellation of $k_{z0}k_{z1}$ and insertion of eqn. (3.37) and $d_{\hat{0}}d_1 \approx 0$:

$$1 + k_z^*(d_{\hat{0}} + d_1) = 0 \quad . \tag{3.42}$$

This equation obviously has no solutions for real $k_z^* > 0$.

The range of validity of the approximate eqn. (3.39) can be found very easily with the help of $k_{\rho p}$-values which can be calculated exactly. Figs. 3.11a and 3.11b show some examples. In all cases eqn. (3.39) produces smaller results for $k_{\rho p}$ than the exact solution, and therefore higher phase velocities. If a relative error is defined as the difference between exact and approximate solution relative to k and a maximum error of 10% is admitted, the range of validity is limited as follows:

$$\varepsilon_{r0} = 2.5 \quad , \varepsilon_{r1} = 10 \quad \begin{cases} d_{\hat{0}} = 0.0 \text{ cm}, & d_1 < 0.18 \text{ cm} \\ d_{\hat{0}} = 0.1 \text{ cm}, & d_1 < 0.15 \text{ cm} \\ d_{\hat{0}} = 0.2 \text{ cm}, & d_1 < 0.12 \text{ cm} \end{cases} \quad . \tag{3.43}$$

The relative error is smaller than 5% if the usual ceramic material is used with $\varepsilon_{r1} \approx 10$ and $d_1 = 0.635$ mm. The range of validity of eqn. (3.39) is larger compared with the data in (3.43) if ε_{r0} and ε_{r1} take on smaller values.

3.2.4.2 Field components of surface waves

The equations for the amplitudes of the surface waves result from the second term of eqn. (2.98) by applying the rule of l'Hospital. One can proceed formally, as is well known, by substituting k_ρ by $k_{\rho p}$ in the numerator and the denominator

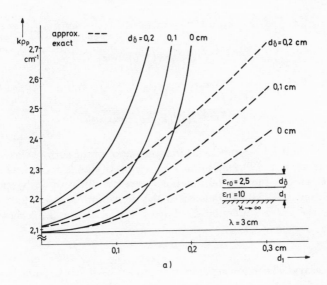

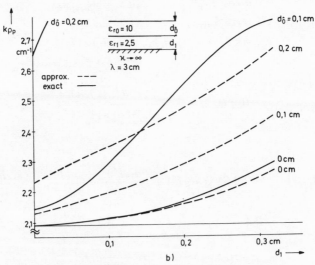

Fig. 3.11 Exact and approximate values of the propagation constants of the TM₁ - surface wave

by its derivative with respect to k_ρ at $k_{\rho p}$ in the integrands of eqns. (3.5) to (3.10). Thus, one gets for example

$$
dE_{\rho 0 w}^{TM} = -2\pi j \frac{B_{0p}}{k_{\rho p}^{TM}} \left[-jk_{zp}\cos(k_{z0p}(d_{\hat{0}} - z)) + \frac{k_{z0p}}{\varepsilon_{r0}}\sin(k_{z0p}(d_{\hat{0}} - z)) \right]
$$
$$
e^{-jk_{zp}d_{\hat{0}}} H_1^{(2)\prime}(k_{\rho p}^{TM}\rho)\cos\varphi \quad , \tag{3.44}
$$

$$
dE_{\rho 0 w}^{TE} = -2\pi j \frac{j\omega\mu_0 D_{0p}}{(k_{\rho p}^{TE})^2 \rho\, k_{z0p}} \left[k_{z0p}\cos(k_{z0p}(d_{\hat{0}} - z)) + jk_{zp}\sin(k_{z0p}(d_{\hat{0}} - z)) \right]
$$
$$
e^{-jk_{zp}d_{\hat{0}}} H_1^{(2)}(k_{\rho p}^{TE}\rho)\cos\varphi \quad . \tag{3.45}
$$

for the E_ρ^{TM}- and E_ρ^{TE}-components in layer "zero" of the surface wave belonging to the pole at $k_{\rho p}$. (The additional subscript w stands for "surface wave".) In these equations the Hankel functions were not substituted by exponential functions (eqn. (2.78)) in contrast to the derivation of eqn. (2.98), because results for the Green's function discussed below in section 3.2.5 show that the restriction

$$
\rho \gg \lambda \tag{3.46}
$$

is not necessary. The following abbreviations are used for the somewhat lengthy expressions for B_{0p} and D_{0p}

$$
C0 = \cos(k_{z0p}d_{\hat{0}}) \quad , \qquad S0 = \sin(k_{z0p}d_{\hat{0}}) \quad , \tag{3.47}
$$
$$
C1 = \cos(k_{z1p}d_1) \quad , \qquad S1 = \sin(k_{z1p}d_1) \quad , \tag{3.48}
$$
$$
K0 = k_{z0p} \, , \qquad K = k_{zp} \, , \qquad K1 = k_{z1p} \quad . \tag{3.49}
$$

Then one gets

$$
B_{0p} = -\frac{I_x dl_x k_{\rho p}^2 S1\, e^{+jKd_{\hat{0}}}\, K0\, K1}{4\pi\omega\varepsilon_{r0}\varepsilon_{r1}\frac{\partial}{\partial k_\rho}(B_{iN})\big|_{k_\rho = k_{\rho p}}} \tag{3.50}
$$

with

$$
\frac{1}{k_\rho}\frac{\partial}{\partial k_\rho}(B_{iN})\Big|_{k_\rho = k_{\rho p}} = S0\, S1 \left[j\frac{d_{\hat{0}}}{\varepsilon_{r1}}K1 + j\frac{d_1}{\varepsilon_{r0}}\frac{K0^2}{K1} + \frac{\varepsilon_{r0}}{\varepsilon_{r1}}\frac{K1}{K} + \frac{\varepsilon_{r0}}{\varepsilon_{r1}}\frac{K}{K1} \right] \tag{3.51}
$$
$$
+ C0\, C1 \left[-j\frac{d_1}{\varepsilon_{r1}}K0 - j\frac{d_{\hat{0}}}{\varepsilon_{r0}}K0 - \frac{K0}{K} - \frac{K}{K0} \right]
$$
$$
+ S0\, C1 \left[-j\frac{2}{\varepsilon_{r0}} + d_{\hat{0}}K + \frac{\varepsilon_{r0}}{\varepsilon_{r1}}d_1 K \right]
$$
$$
+ S1\, C0 \left[-j\frac{1}{\varepsilon_{r1}}\frac{K1}{K0} - j\frac{1}{\varepsilon_{r1}}\frac{K0}{K1} + \frac{\varepsilon_{r0}}{\varepsilon_{r1}}d_{\hat{0}}\frac{K\,K1}{K0} + d_1\frac{K\,K0}{K1} \right]
$$

and

$$D_{0p} = -\frac{I_x dl_x k_{\rho p}^2}{4\pi} \frac{S1 \, e^{jK d_{\hat{0}}} \, K0}{\frac{\partial}{\partial k_\rho}(D_{\hat{1}N})\big|_{k_\rho = k_{\rho p}}} \qquad (3.52)$$

with

$$\frac{1}{k_\rho}\frac{\partial}{\partial k_\rho}(D_{\hat{1}N})\big|_{k_\rho = k_{\rho p}} = S0\,S1\,[j d_1 K + 2 + j d_{\hat{0}} K] \qquad (3.53)$$

$$+ S0\,C1\left[d_{\hat{0}}K1 - j\frac{K}{K1} - j\frac{K1}{K} + d_1\frac{K0^2}{K1}\right]$$

$$+ C0\,S1\left[d_1 K0 + d_{\hat{0}}K0 - j\frac{K0}{K} - j\frac{K}{K0}\right]$$

$$+ C0\,C1\left[-\frac{K1}{K0} - \frac{K0}{K1} - j d_{\hat{0}}\frac{K\,K1}{K0} - j d_1\frac{K\,K0}{K1}\right] \quad .$$

One can see from these equations that the surface waves - as shown above in section 3.2.3 - propagate radially with a phase velocity

$$v_{ph}^{TM/TE} = \frac{\omega}{k_{\rho p}^{TM/TE}} \qquad (3.54)$$

(assuming that all materials are lossfree and thus $k_{\rho p}^{TM/TE}$ real). In the area $\hat{1}$, i.e. above the layers, the amplitudes decay exponentially with z because of

$$k_{zp} = \sqrt{k^2 - k_{\rho p}^2}\,, \qquad k_{\rho p} > k \quad , \qquad (3.55)$$

implying that the wave is guided by the layers. Within the layered structure the field components are described by trigonometric functions. If $k_{\rho p}$ is larger than k_0 or k_1 one obtains an exponential dependence on z within the corresponding layer as well.

One readily sees by means of eqns. (A3.47) to (A3.58) and eqn. (2.78) that the electric field components $E_{\rho w}^{TM}$, E_{zw}^{TE} and $E_{\varphi w}^{TE}$ are proportional to $\rho^{-1/2}$ in the farfield and that $E_{\varphi w}^{TM}$ and $E_{\rho w}^{TE}$ are proportional to $\rho^{-3/2}$. Thus, for $\rho \gg \lambda$, $E_{\varphi w}^{TM}$ and $E_{\rho w}^{TE}$ can be neglected - this is the case we have discussed in the context of the saddle-point solution. Thus the electric field of a TM-surface wave essentially only has a ρ and a z-component, the field of a TE-surface wave only has a φ-component.

Since $E_{zw}^{TM} \sim \cos(\varphi)$, $E_{\rho w}^{TM} \sim \cos(\varphi)$ and $E_{\varphi w}^{TE} \sim \sin(\varphi)$ a TM-surface wave propagates mainly in the direction of the exciting current. The TE-surface wave propagates perpendicular to the exciting current.

Similarly to the discussion of propagation constants above, the equations for the field components of the TM-surface wave can be enormously simplified with the assumption that $d_{\hat{0}} \ll \lambda_0$ and $d_1 \ll \lambda_1$. Multiplication of the denominator of $B_{\hat{1}}$ (eqn. (3.27)) with k_{z1} and differentiation with respect to k_ρ yields, if terms containing $d_1 d_{\hat{0}}$ are neglected

$$\frac{\partial}{\partial k_\rho}(k_{z1}B_{\hat{1}N}) \approx -\left(1 + 2k_z(j\frac{1}{\varepsilon_{r1}}d_1 + j\frac{1}{\varepsilon_{r0}}d_{\hat{0}})\right)\frac{k_\rho}{k_z} \approx -\frac{k_\rho}{k_z} \quad . \tag{3.56}$$

Furthermore, because $k_{\rho 1p}^{TM} \approx k$, we set according to eqn. (3.40)

$$k_{z1p} \approx \sqrt{k^2(\varepsilon_{r1} - 1) + k^2 - k_{\rho 1p}^2} \approx k\sqrt{\varepsilon_{r1} - 1} \quad . \tag{3.57}$$

Then one obtains the following equations for the electric field in space $\hat{1}$, i.e. above the layers,

$$dE_{zi w}^{TM_1} \approx \frac{I_x dl_x}{\sqrt{2\pi}} \sqrt{\frac{\mu_0}{\varepsilon_0}} k^2 \frac{\varepsilon_{r1} - 1}{\varepsilon_{r1}} d_1(d_1 + d_{\hat{0}})(1 - \frac{1}{\varepsilon_{eff}})k_{\rho 1p}^2 \tag{3.58}$$

$$\frac{e^{-j(k_{z1p}z + k_{\rho 1p}\rho - \frac{3}{4}\pi)}}{\sqrt{k_{\rho 1p}\rho}} \cos\varphi$$

and

$$dE_{\rho \hat{1}w}^{TM_1} \approx -\frac{k_{z1p}}{k_{\rho 1p}} dE_{z\hat{1}w}^{TM_1} \quad . \tag{3.59}$$

Figures 3.12a - 3.12d show the field components tangential to the interfaces for the excited surface waves in the plane of the exciting current ($z = 0$) for the same structures as in the previous section. We have drawn the modulus of $dE_{\rho \hat{1}w}$ and $dE_{\varphi \hat{1}w}$ normalized by $\cos\varphi$ and $\sin\varphi$ separately. All examples demonstrate that the excitation of the TM_1-surface wave is low for thin layers ($d_{\hat{0}} < 1$ mm). For the usual ceramic substrate ($\varepsilon_{r1} = 10, d_1 = 0.635$ mm without cover) for example, the value of $\mid dE_{\rho \hat{1}w} \mid$ is only 0.4% of the corresponding value for a substrate thickness of 2.55 mm. If $\varepsilon_{r1} = 10$ and $\varepsilon_{r0} = 2.5$ (Fig. 3.12a), then first of all the amplitude of the TM_1- surface wave increases with the thickness of the substrate for fixed thickness of the cover. It falls as soon as the TE_1-surface wave can propagate. If the thickness of the substrate increases even more, the curves for the amplitudes of the higher TM- and TE-surface waves also pass through maxima. The height of these maxima falls distinctly with increasing thickness of the cover. In the case of Fig. 3.12b the dielectric constant of the cover substantially exceeds that of the substrate. Now mainly the TE_1-surface wave is excited for $d_1 > 5$ mm and the amplitudes of all propagating modes are nearly independent of d_1 for thick substrates. For all the results presented in Figs. 3.12c and 3.12d the dielectric constants of the substrate and the cover material are assumed to be identical. The graphs for $\varepsilon_{r1} = 2.5$ resemble those

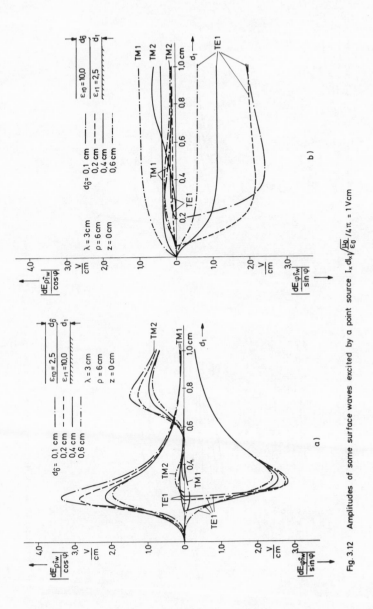

Fig. 3.12   Amplitudes of some surface waves excited by a point source $I_x\,d_x\sqrt{\dfrac{\mu_0}{\varepsilon_0}}/4\pi = 1\,\mathrm{Vcm}$

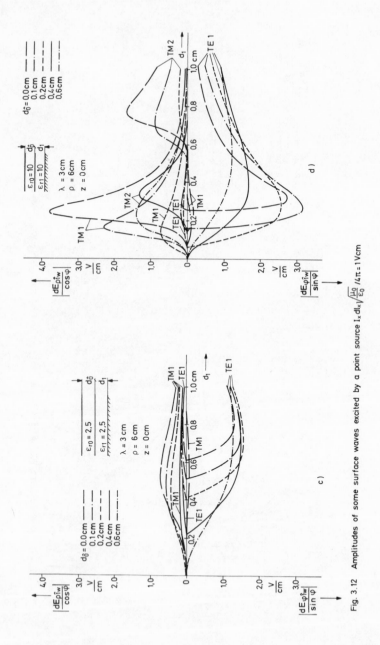

Fig. 3.12   Amplitudes of some surface waves excited by a point source $I_x \, dl_x \sqrt{\frac{\mu_0}{\varepsilon_0}} / 4\pi = 1 Vcm$

of Fig. 3.12b, the graphs for $\varepsilon_{r1} = 10$ those of Fig. 3.12a. Their shapes confirm
the obvious assumption that the amplitudes of the surface waves are not only
determined by the dielectric constants of the materials used but also by the
position of the exciting current source within the layers. One therefore gets for
example from Fig. 3.12d for

$$d_{\hat{0}} = 0.1 \text{ cm} \quad \text{and} \quad d_1 = 0.2 \text{ cm} \qquad |\frac{dE_{\rho 1 w}^{TM_1}}{\cos\varphi}| = 2.0\frac{\text{V}}{\text{cm}} \qquad (3.60)$$

and for

$$d_{\hat{0}} = 0.2 \text{ cm} \quad \text{and} \quad d_1 = 0.1 \text{ cm} \qquad |\frac{dE_{\rho 1 w}^{TM_1}}{\cos\varphi}| = 0.6\frac{\text{V}}{\text{cm}} \quad . \qquad (3.61)$$

The influence of icing ($\varepsilon_{r0} = 3.17 - \text{j}0.0025$) is similar to that of a Teflon cover
($\varepsilon_r = 2.5$).

We compare exact and approximate values for the discussion of the approximate equation (3.58). As can be concluded from Figs. 3.13a,b and from many additional results it only makes sense to use eqn. (3.58) if the cover is very thin, for example for paint coatings. As an example one gets the following relative errors:

ε_{r1}	ε_{r0}	d_1	$d_{\hat{0}}$	error
2.5	1.0	0.1 cm	0	under 1%
2.5	2.5	0.1 cm	0.1 cm	cir. 100%
10.0	2.5	0.1 cm	0	cir. 23%

$$\qquad (3.62)$$

If we set $d_{\hat{0}} = 0.1$ cm in the last example and vary d_1, then for increasing thickness of the substrate the approximate equation first of all yields too large values, and from $d \approx 0.1$ cm onwards too small values (Fig. 3.13a). For cover layers with a thickness of 2 mm the curve resulting from eqn. (3.58) can look completely different from the exact curve, cf. (Fig. 3.13b).

3.2.5 Space wave

Starting from eqns. (3.5) and (3.6) one can determine the field of the space wave in spherical coordinates (r, ϑ, φ) by applying eqn. (2.98). For this the results for $dE_{\rho s}^{TM}$ and dE_{zs}^{TM} are combined to $dE_{\vartheta s}^{TM}$, as in eqn. (2.105). The final result reads:

$$dE_{\varphi s}^{TE} = j\frac{I_x dl_x}{4\pi}\omega\mu_0 \sin\varphi \frac{e^{-jkr}}{r}f_\varphi(\vartheta) \quad , \qquad (3.63)$$

$$dE_{\vartheta s}^{TM} = j\frac{-I_x dl_x}{4\pi}\omega\mu_0 \cos\varphi \frac{e^{-jkr}}{r}f_\vartheta(\vartheta) \qquad (3.64)$$

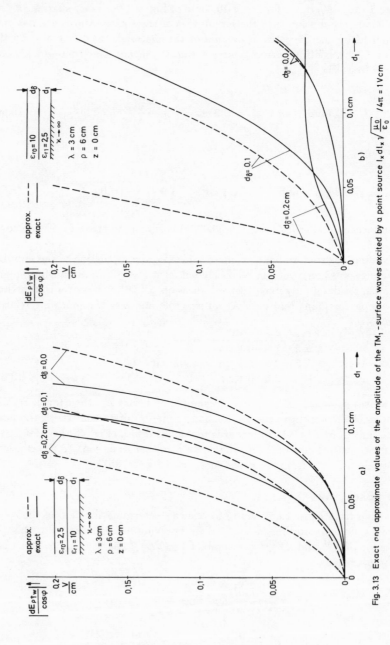

Fig. 3.13   Exact and approximate values of the amplitude of the TM$_1$ -surface waves excited by a point source $I_x dI_x \sqrt{\frac{\mu_0}{\epsilon_0}} / 4\pi = 1 Vcm$

with

$$f_\varphi(\vartheta) = \frac{2\eta_0 \cos\vartheta \left[\cos(\,d_{\hat 0} k \cos\vartheta\,) + j \sin(\,d_{\hat 0} k \cos\vartheta\,)\right]}{\cos\vartheta \left[\eta_0 \cos(\,d_{\hat 0} k\eta_0) + \eta_1 \sin(\,d_{\hat 0} k\eta_0) \operatorname{ctg}(kd_1\eta_1)\right]} \tag{3.65}$$

$$\overline{\qquad -j\eta_0[\eta_1 \cos(\,\eta_0 kd_{\hat 0})\operatorname{ctg}(\eta_1 kd_1) - \eta_0 \sin(\,\eta_0 kd_{\hat 0})] \qquad} \quad,$$

$$f_\vartheta(\vartheta) = \frac{2\eta_0\eta_1\varepsilon_{r0} \cos\vartheta \left[\cos(\,d_{\hat 0} k \cos\vartheta\,) + j \sin(\,d_{\hat 0} k \cos\vartheta\,)\right]}{\eta_0[\varepsilon_{r0}\eta_1 \cos(\,\eta_0 kd_{\hat 0}) + \varepsilon_{r1}\eta_0 \sin(\,\eta_0 kd_{\hat 0}) \operatorname{ctg}(\eta_1 kd_1)]} \tag{3.66}$$

$$\overline{\qquad -j\varepsilon_{r0} \cos\vartheta \left[\varepsilon_{r1}\eta_0 \cos(\,\eta_0 kd_{\hat 0}) \operatorname{ctg}(\eta_1 kd_1) - \varepsilon_{r0}\eta_1 \sin(\,\eta_0 kd_{\hat 0})\right] \qquad} \quad,$$

$$\eta_0 = \sqrt{\varepsilon_{r0} - \sin^2 \vartheta} \quad, \tag{3.67}$$

$$\eta_1 = \sqrt{\varepsilon_{r1} - \sin^2 \vartheta} \quad. \tag{3.68}$$

Eqns. (3.65) to (3.66) can be simplified for the case of thin layers as well. For $\eta_0 kd_{\hat 0} \ll 1$ and $\eta_1 kd_1 \ll 1$ we get:

$$f_\varphi(\vartheta) \approx \frac{2kd_1 \cos\vartheta}{k(d_{\hat 0} + d_1)\cos\vartheta - j} \quad, \tag{3.69}$$

$$f_\vartheta(\vartheta) \approx \frac{2kd_1\varepsilon_{r0}\eta_1^2 \cos\vartheta}{k[\varepsilon_{r0}\eta_1^2 d_1 + \varepsilon_{r1}\eta_0^2 d_{\hat 0}] - j\varepsilon_{r0}\varepsilon_{r1} \cos\vartheta} \quad. \tag{3.70}$$

In eqns. (3.65) and (3.66) the thicknesses of the layers $d_{\hat 0}$ and d_1 are contained in the argument of trigonometric functions. Thus, the results for f_φ and f_ϑ strongly depend on the values for $d_{\hat 0}$ and d_1 especially if the dielectric constants ε_{r0} and ε_{r1} are large. Because of the multitude of possible combinations of $\varepsilon_{r1}, \varepsilon_{r0}, d_1$ and $d_{\hat 0}$ a systematic discussion hardly appears to be feasible. That is why in Fig. 3.14 only the thickness of the substrate d_1 has been varied while leaving $\varepsilon_{r0}, \varepsilon_{r1}$ and $d_{\hat 0}$ unchanged. Note that by modifying just this single parameter one can achieve very different results for the radiation pattern of the space wave.

3.2.6 Nearfield of a point source in the plane $z = 0$

The integration is carried out numerically for observation points in the nearfield ($\rho \le 2\lambda$) as mentioned in section 2.4.3. Due to the poles on the real axis between k and $\max\{k_0, k_1\}$ the integration is carried out in three sections

$$I \qquad k_\rho' < k \quad, \tag{3.71}$$

$$II \qquad k \le k_\rho' \le \max\{k_0, k_1\} \quad, \tag{3.72}$$

$$III \qquad k_\rho' > \max\{k_0, k_1\} \quad. \tag{3.73}$$

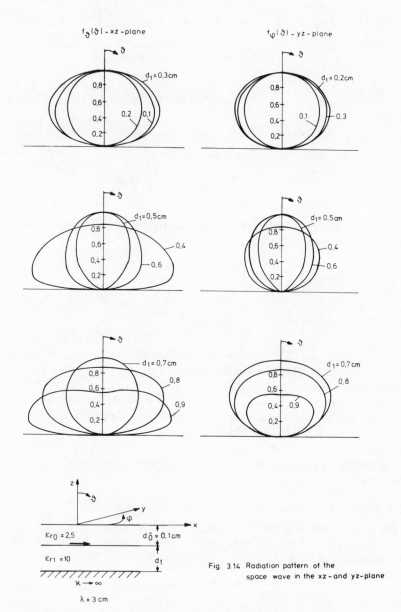

Fig. 3.14 Radiation pattern of the
space wave in the xz- and yz-plane

Hence it is possible to choose the width of the intervals for Simpson's rule separately for each of the three sections. One has to make sure, for example, that there are enough intervals between two poles at close distance.

Figs. 3.15a - 3.15d present the graphs of the tangential field dE_ρ and dE_φ in a plane immediately above the source ($z = 0.003$ cm) for a wavelength of 3 cm for some structures discussed previously. The parameters for the first two figures are chosen in such a way that only the TM_1-surface wave can propagate. Its curve according to the saddle-point solution (eqn. (3.44)) has been drawn in addition. The range of validity of the saddle-point solution is restricted by two conditions according to section 2.4.2: poles must not be too close to the saddle point and the observation point must be far enough away from the source point. Regarding the present results one can see that these conditions do not impose severe restrictions. In Fig. 3.15a for example

$$k_{\rho p}^{TM_1} = 2.17 \frac{1}{\text{cm}} = 1.04\,k \quad . \tag{3.74}$$

Hence

$$w_p = \frac{\pi}{2} + j\,0.28 \quad , \tag{3.75}$$

and the saddle point is located at

$$w_s = \vartheta = \tan^{-1}\frac{\rho}{z} \quad . \tag{3.76}$$

Although pole and saddle point are close together for $z = 0.003$ cm and $\rho \gtrsim 0.03$ cm, eqn. (3.44) describes the behaviour of dE_ρ^{TM} very well if the observation point is farther away from the source point than one wavelength. In this range the amplitude of the E_φ- component is substantially smaller than that of the E_ρ-component, as has been derived in section 3.2.4.2. Considering the phases in Figs. 3.15a and 3.15b one notes that the E_φ-component, too, is determined by the surface wave: in the case of observation points with $\rho \geq \lambda$ the curves of $\arg(dE_\rho^{TM1})$ and $\arg(dE_\varphi^{TM1})$ are almost parallel. If the TM_1-wave is strongly excited - as is the case in Fig. 3.15b - the approximate equations for the surface waves describe the field very well for all observation points with $\rho > 0.5\lambda$. In the structure according to Fig. 3.15c both the TM_1-wave and the TE_1-wave can propagate, the amplitudes being roughly equal. The enormously different propagation constants of these waves can be clearly seen by considering the curves for the phases.

If several surface waves are excited one gets an oscillating curve for the amplitudes of both field components, cf. Fig. 3.15d.

Starting from eqns. (2.169) to (2.173) in section 2.4.4.2 one can also derive simple approximate equations which are valid for observation points very close to the source. For this we set $\varepsilon_{r2} = \varepsilon_{r3}$ in eqn. (2.173) and then the multiplication explained in subsection b) is carried out. If only the images of the source

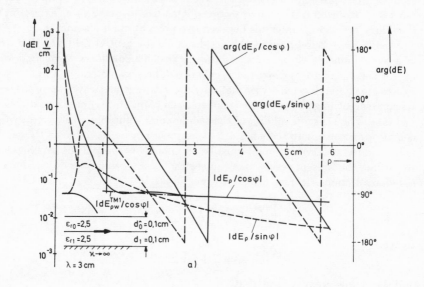

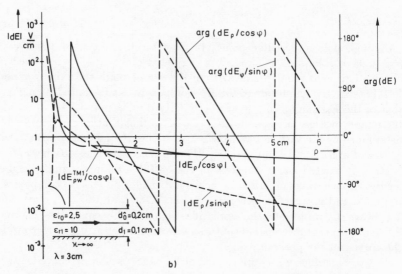

Fig. 3.15 Electric field in the plane z = 0,003 cm excited by a point source $I_x dl_x \sqrt{\mu_0/\epsilon_0}/4 = 1Vcm$

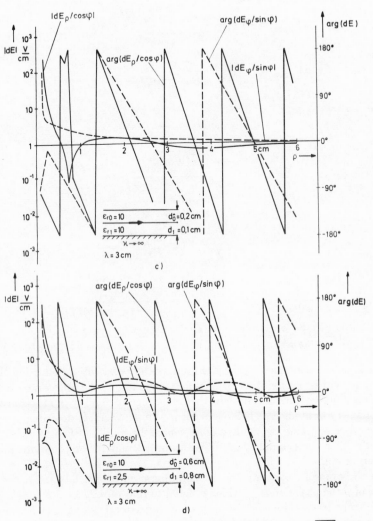

Fig. 3.15 Electric field in the plane $z = 0.003\,cm$ excited by a point source $I_x\,dl_x\sqrt{\mu_0/\varepsilon_0}/4\pi = 1\,Vcm$

produced by the reflector and by the interface cover-free space are taken into account, this results in the following equations for the dE_ρ and dE_φ component in layer "$\hat{0}$" (which is equivalent to layer "1" in section 2.4.4.2):

$$dE_{\rho\hat{0}\infty} \approx -j\frac{I_x dl_x}{4\pi}\frac{\cos\varphi}{\omega\varepsilon_0}\sum_{q=0}^{3}\frac{a_q}{r_q^3}(3\sin^2\vartheta_q - 1) \quad , \tag{3.77}$$

$$dE_{\varphi\hat{0}\infty} \approx -j\frac{I_x dl_x}{4\pi}\frac{\sin\varphi}{\omega\varepsilon_0}\sum_{q=0}^{3}\frac{a_q}{r_q^3} \tag{3.78}$$

with

$$\sin\vartheta_q = \frac{\rho}{r_q} \quad , \tag{3.79}$$

$$r_0 = r = \sqrt{\rho^2 + z^2} \quad , \tag{3.80}$$

$$r_1 = \sqrt{\rho^2 + (z + 2d_1)^2} \quad , \tag{3.81}$$

$$r_2 = \sqrt{\rho^2 + (z + 2d_{\hat{0}})^2} \quad , \tag{3.82}$$

$$r_3 = \sqrt{\rho^2 + (z - 2d_{\hat{0}})^2} \quad , \tag{3.83}$$

and

$$a_0 = \frac{2}{\varepsilon_{r0} + \varepsilon_{r1}} \quad , \qquad\qquad a_1 = -\frac{4\varepsilon_{r1}}{(\varepsilon_{r0} + \varepsilon_{r1})^2} \quad , \tag{3.84}$$

$$a_2 = -\frac{2(\varepsilon_{r0} - \varepsilon_{r1})(1 - \varepsilon_{r0})}{(\varepsilon_{r0} + \varepsilon_{r1})^2(1 + \varepsilon_{r0})} \quad , \qquad a_3 = -\frac{2(1 - \varepsilon_{r0})}{(\varepsilon_{r0} + \varepsilon_{r1})(1 + \varepsilon_{r0})} \quad . \tag{3.85}$$

In Fig. 3.16 some fields calculated numerically and calculated using the approximate equations for observation points very close to the source are compared with each other. The curves show that the range of validity of the approximations is very limited. The behaviour for $\rho \to 0$, however, is described correctly. Since, when applying the method of moments, the coupling of currents very near to each other (for example 0.001λ) has to be calculated, the equations prove to be very useful.

In the examples shown $|dE_\varphi|$ has a minimum at $\rho \simeq 0.06\lambda$. At the same value of ρ the phase of dE_φ, which is equal to that of dE_ρ for $\rho \lesssim 0.04\lambda$, jumps by roughly 180°. This effect can now be understood taking into account our discussion of the nearfield and the surface waves: the phases of dE_ρ and dE_φ are equal in observation points very close to the source, according to eqns. (3.77) and (3.78), whereas they differ in the case of surface waves, as can be seen, e.g. in Figs. 3.15a and 3.15b. The minimum occurs roughly at the point where the contributions of the nearfield and of the surface wave roughly compensate each other.

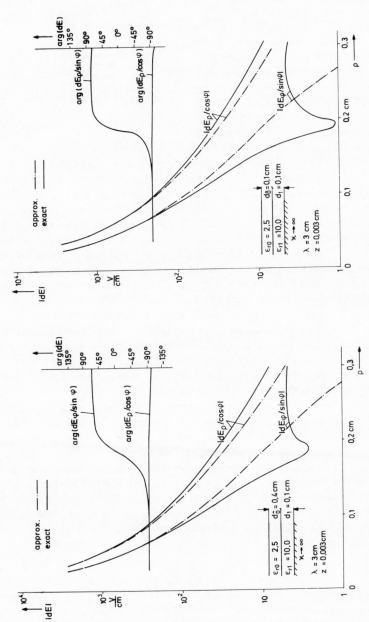

Fig. 3.16 Exact and approximate values of the electric field excited by a point source $I_x dl_x \sqrt{\frac{\mu_0}{\varepsilon_0}}/4\pi = 1\,Vcm$

3.3 Dipoles of finite lengths

3.3.1 Current distribution and input impedance

To start with a simple example, the radiation property of a thin dipole of finite length which is fed in the centre by a voltage-source (Fig. 3.17) shall be determined. This is based on the Green's functions, calculated before and shown for example in Fig. 3.15. Because of

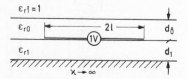

Fig. 3.17 Printed dipole with a dielectric cover

the singularity of the Green's function for ρ approaching zero, only the elements on the diagonal Z_{ii} and the two neighbouring elements $Z_{i\pm1,i}, Z_{i\pm2,i}$ of the system of linear equations (A1.4) are substantially different from zero. According to eqns. (A1.10) and (3.77) to (3.85) magnitudes and phases of these matrix elements are mainly determined by the first term of the sums in eqns. (3.77) and (3.78). Apart from the coordinates of the observation point, this term only contains the dielectric constants ε_{r0} and ε_{r1}, but not the thicknesses of the layers and ε_{ri}. One expects therefore that the current distribution on the dipole is above all determined by the dielectric constants of the layers "0" and "1" which are in immediate contact with the dipole. This is confirmed by many examples calculated by the authors. Figs. 3.18 and 3.19 show some examples for substrates and covers with ε_r=2.5 and ε_r=10, each with different dipole lengths and thicknesses of layers. The voltage is 1 V in all cases at a frequency of 10 GHz. For the results of Fig. 3.19, the parameters of substrate and cover have been interchanged compared to those of Fig. 3.18. A comparison of the two diagrams reveals that $|I(x)| / |I(0)|$, i.e. the relative current distribution, depends effectively only on the lengths of the dipoles. If the length is $l \simeq 2.7$ mm the current distribution corresponds to that of a $\lambda_e/2$-resonator, at $l \simeq 5.1$ mm it corresponds to that of a λ_e-resonator. Up to a dipole length of about 3.0 mm the phase of the current on the dipole is almost constant, which could be demonstrated by a lot of additional calculations. When dealing with larger dipole lengths a large deviation from this constant behaviour could be observed (for example Figs. 3.18 and 3.19, $l = 5.1$ mm). This fact is not surprising because the current distribution on a dipole can roughly be interpreted as that of a standing wave on a transmission line, which has a phase jump of π at a distance of $\lambda_e/2$ from the open end of the line. It should be added that the computer program developed by the authors is only applicable if

$$d_{\hat{0}/1} \gtrsim \frac{1}{20} \frac{\lambda}{\sqrt{\varepsilon_{r0/1}}} \tag{3.86}$$

holds for the layers. The case $d_{\hat{0}} = 0$, i.e. no cover, can also be treated as well, of course.

The dependence of the current amplitudes in the current maximum and also in the feeding point becomes visible in the diagram for the input impedance as

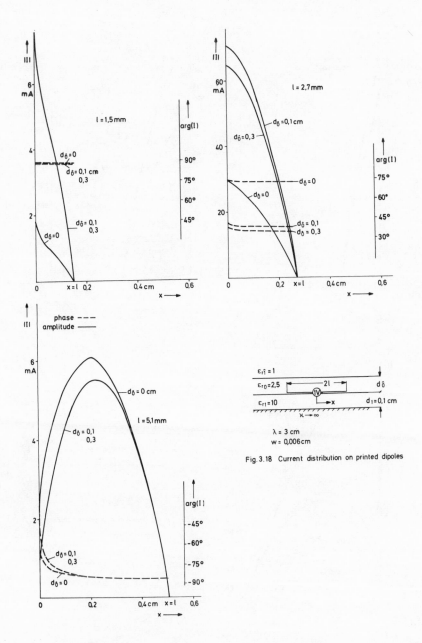

Fig.3.18 Current distribution on printed dipoles

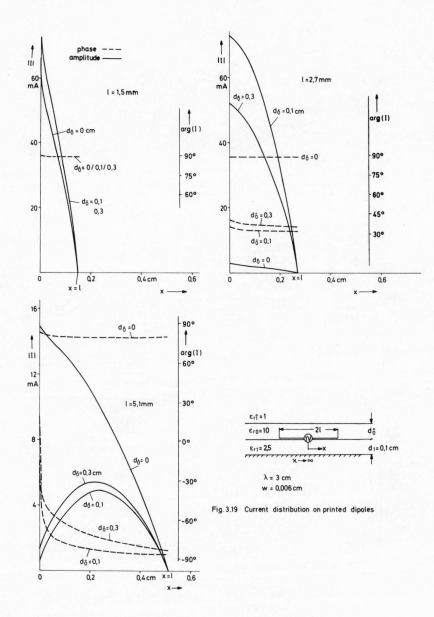

Fig. 3.19 Current distribution on printed dipoles

a function of the dipole length (Fig. 3.20). In order to illustrate the behaviour at the first resonance point, this area has been enlarged. Note that the cover above all causes a shift of the curve down to smaller dipole lengths whereas the real part of the input impedance in the first resonance point remains roughly unchanged. The position of this first resonance point changes only slightly for thicknesses of

$$d_{\hat{0}/1} \gtrsim 1 \text{ mm} \quad . \tag{3.87}$$

The influence on the value of the real part in the second resonance point and its position is stronger. From Fig. 3.20b for example the resonance lengths can be read off:

$$l_{\text{res1}}(d_{\hat{0}} = 0.1 \text{ cm}) \approx 2.80 \text{ mm} \quad , \tag{3.88}$$

$$l_{\text{res1}}(d_{\hat{0}} = 0.3 \text{ cm}) \approx 2.85 \text{ mm} \quad , \tag{3.89}$$

$$l_{\text{res2}}(d_{\hat{0}} = 0.1 \text{ cm}) \approx 4.95 \text{ mm} \quad , \tag{3.90}$$

$$l_{\text{res2}}(d_{\hat{0}} = 0.3 \text{ cm}) \approx 5.20 \text{ mm} \quad . \tag{3.91}$$

If the dipoles were situated in a homogeneous space with a dielectric constant

$$\frac{\varepsilon_{r0} + \varepsilon_{r1}}{2} = 6.25 \quad , \tag{3.92}$$

one would get the resonance lengths

$$l_{res1} \approx 2.84 \text{ mm} \quad , \tag{3.93}$$

$$l_{res2} \approx 5.06 \text{ mm} \quad . \tag{3.94}$$

The position of the first resonance point is therefore, just like the relative current distribution, effectively determined by the materials of the immediate surroundings of the dipole and by its length and width. This result shows that it must be possible to find simple approximate equations for the current distribution and the input impedance. On the other hand, it becomes evident that one can hardly obtain data of the setup of a stratified structure by measuring the input impedance of a dipole in this structure.

3.3.2 Radiation pattern

Knowing the current distribution one is able to calculate the radiation pattern with the help of eqns. (3.63) to (3.68). In the present case of a straight dipole the current only has an x-component. The radiation field must therefore be calculated according to the equation

$$\vec{E} = j\frac{\omega\mu_0}{4\pi}\frac{e^{-jkr}}{r}\left(\vec{u}_\varphi \sin\varphi \, f_\varphi(\vartheta) - \vec{u}_\vartheta \cos\varphi \, f_\vartheta(\vartheta)\right)F(\vartheta,\varphi) \tag{3.95}$$

Printed antennas

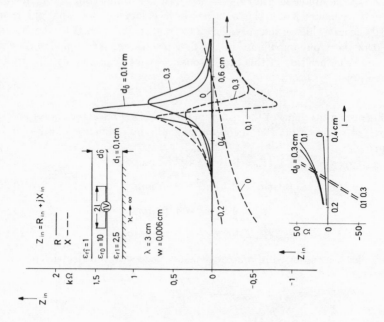

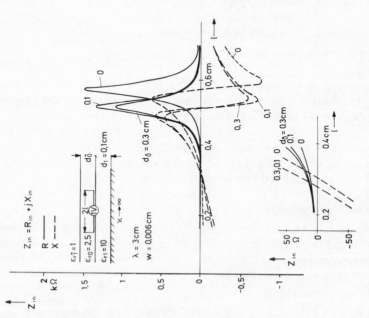

Fig. 3.20 Input impedance of printed dipoles

with

$$F(\vartheta, \varphi) = \int\limits_{2l} I_x(x') e^{jk\vec{u}_r\vec{r}'} dx' \quad .$$ (3.96)

The primed variables x' and $\vec{r}'$ again denote the position of the source point, $\vec{u}_r r$ denotes the position of the observation point. Following the results for the current distribution discussed in the previous section we can assume a sinusoidal current distribution as a first approximation:

$$I_x \approx I_x(0) \sin\left(\frac{2\pi}{\lambda_e}(l - |x|)\right)$$ (3.97)

with

$$\frac{2\pi}{\lambda_e} = k_e = \omega\sqrt{\mu_0\varepsilon_0\varepsilon_{re}}$$ (3.98)

and

$$\varepsilon_{re} = \frac{\varepsilon_{r0} + \varepsilon_{r1}}{2} \quad .$$ (3.99)

Then eqn. (3.96) can be integrated analytically, and for a dipole of the length $\lambda_e/2$ one gets

$$F(\vartheta, \varphi) \approx \frac{2}{k_e} \frac{\cos\left(\frac{\pi}{2}\frac{\sin\vartheta}{\sqrt{\varepsilon_{re}}}\cos\varphi\right)}{1 - \frac{1}{\varepsilon_{re}}(\sin\vartheta\,\cos\varphi)^2} I_x(0) \quad .$$ (3.100)

If ε_{re} is much larger than 1, then

$$F(\vartheta, \varphi) \approx \frac{2}{k_e} I_x(0) \quad .$$ (3.101)

For typical ceramic substrates with $\varepsilon_{r1} = 10$ and

$$\lambda_e = \frac{\lambda}{\sqrt{\varepsilon_{re}}} \approx \frac{\lambda}{3}$$ (3.102)

eqn. (3.101) is reasonably good in many cases of practical importance. Generally one finds that in the case of short dipoles the term $e^{jk\vec{u}_r\vec{r}'}$ in eqn. (3.96) can be replaced by 1. Then F is no longer a function of the angles ϑ and φ and the radiation field according to eqn. (3.95) is only determined by the Green's function and not by the current distribution.

The total power radiated by the space wave can be calculated by integrating over the radiation field

$$P_{sr} = \int\limits_{\vartheta=0}^{\pi/2} \int\limits_{\varphi=0}^{2\pi} \frac{1}{2} \frac{|E_\vartheta|^2 + |E_\varphi|^2}{\sqrt{\frac{\mu_0}{\varepsilon_0}}} r^2 \sin\vartheta\, d\varphi d\vartheta \quad .$$ (3.103)

In general this integral can only be evaluated numerically. An analytical solution is obtained for the case of thin layers and large $\varepsilon_{r0/1}$. Using eqns. (3.69), (3.70) and (3.101) one gets, for a dipole of length $\lambda_e/2$ and a current distribution according to eqn. (3.97):

$$P_{sr} \approx \frac{1}{2} \frac{\sqrt{\frac{\mu_0}{\varepsilon_0}}}{\varepsilon_{re}\pi}(kd_1)^2 I_x^2(0) \left\{ \frac{4}{3} - k(d_{\hat{0}} + d_1) \left[\frac{\pi}{2} - k(d_{\hat{0}} + d_1) \right] \right\} \quad . \quad (3.104)$$

Neglecting the second term in curly brackets one obtains the following expression as a first approximation for the power radiated by the space wave

$$P_{sr} \approx \frac{1}{2} \sqrt{\frac{\mu_0}{\varepsilon_0}} \frac{I_x^2(0)}{\varepsilon_{re}\pi}(kd_1)^2 \frac{4}{3} \quad . \quad (3.105)$$

As a comparison we shall determine the power transported by the surface waves. As the dependence on φ is described by $\cos\varphi$ or $\sin\varphi$ and the dependence on z is also given by simple trigonometric functions, corresponding equations can be developed from eqns. (3.5) to (3.10). In the case of thin layers, using eqn. (3.58) one gets the simple expression

$$P_{sw} \approx \frac{1}{2} \frac{\sqrt{\frac{\mu_0}{\varepsilon_0}}}{\varepsilon_{re}} \left(k\,d_1 \frac{\varepsilon_{r1} - 1}{\varepsilon_{r1}} \right)^2 \left(k(d_{\hat{0}} + d_1)(1 - \frac{1}{\varepsilon_{reff}}) \right) I_x^2(0) \quad . \quad (3.106)$$

for the power carried by a TM_1-surface wave which is excited by a $\lambda_e/2$-dipole with a current distribution according to eqn. (3.97). If $d_{\hat{0}} = 0$, i.e. no cover, then $\varepsilon_{re} = \varepsilon_{r1}$ (eqns. (3.39) and (3.40)) and hence

$$P_{sw} \approx \frac{1}{2} \sqrt{\frac{\mu_0}{\varepsilon_0}} \frac{I_x^2(0)}{\varepsilon_{re}} \left(k\,d_1 \frac{\varepsilon_{r1} - 1}{\varepsilon_{r1}} \right)^3 \quad . \quad (3.107)$$

The power transported by the surface wave thus grows with d_1^3, that of the space wave only with d_1^2. Obviously the former can not be disregarded especially if the power carried by the surface wave can actually be radiated at the edges of a substrate of finite size.

In this section two additional properties of the radiation field will be discussed for the case of dielectric constants being substantially larger than 1 ($\varepsilon_{r0/1} \gg 1$). Here the current distribution $I_x(x)$ can be arbitrary.

1) If $\varepsilon_{r0} = \varepsilon_{r1} = \varepsilon_r$, i.e. if the dipole is situated in a homogeneous layer, the radiation pattern does not depend on the individual values of $d_{\hat{0}}$ or d_1 but only on the total thickness of the layer $(d_{\hat{0}} + d_1)$. To prove this statement the products $f(\vartheta) \cdot F(\vartheta, \varphi = \text{const.})$ and $f_\varphi(\vartheta) \cdot F(\vartheta, \varphi = \text{const.})$ from eqn. (3.95)

will be studied which have been normalized by the field strength in the main lobe $\vartheta = \varphi = 0°$. If these terms are called $F_\vartheta(\vartheta, \varphi)$ and $F_\varphi(\vartheta, \varphi)$ and if

$$\varepsilon_{r0/1} \gg \sin^2 \vartheta \tag{3.108}$$

is inserted one gets with the help of eqns. (3.65) to (3.68)

$$
\begin{aligned}
F_\vartheta(\vartheta, \varphi) &= \frac{f_\vartheta(\vartheta) F(\vartheta, \varphi)}{f_\vartheta(0) F(0,0)} \\
&= \frac{[\sin(k\sqrt{\varepsilon_r}\,(d_{\hat{0}} + d_1)) - j\sqrt{\varepsilon_r}\,\cos(k\sqrt{\varepsilon_r}\,(d_{\hat{0}} + d_1))]}{\eta \sin(k\eta(d_{\hat{0}} + d_1)) - j\varepsilon_r \cos\vartheta\,\cos(k\eta(d_{\hat{0}} + d_1))} \\
&\quad \cdot \frac{\eta \cos\vartheta\,\sin(\eta k d_1)}{\sin(\sqrt{\varepsilon_r}\,k d_1)}\, e^{jkd_{\hat{0}}(\cos\vartheta\,-1)} \frac{F(\vartheta, \varphi)}{F(0,0)} \quad ,
\end{aligned}
\tag{3.109}
$$

$$
\begin{aligned}
F_\varphi(\vartheta, \varphi) &= \frac{f_\varphi(\vartheta) F(\vartheta, \varphi)}{f_\varphi(0) F(0,0)} \\
&= \frac{[\sin(k\sqrt{\varepsilon_r}\,(d_{\hat{0}} + d_1)) - j\sqrt{\varepsilon_r}\,\cos(k\sqrt{\varepsilon_r}\,(d_{\hat{0}} + d_1))]}{\cos\vartheta\,\sin(k\eta(d_{\hat{0}} + d_1)) - j\eta \cos(k\eta(d_{\hat{0}} + d_1))} \\
&\quad \cdot \frac{\cos\vartheta\,\sin(\eta k d_1)}{\sin(\sqrt{\varepsilon_r}\,k d_1)}\, e^{jkd_{\hat{0}}(\cos\vartheta\,-1)} \frac{F(\vartheta, \varphi)}{F(0,0)} \quad .
\end{aligned}
\tag{3.110}
$$

Thereby according to our supposition we have

$$\varepsilon_r = \varepsilon_{r0} = \varepsilon_{r1} \tag{3.111}$$

and

$$\eta = \eta_0 = \eta_1 = \sqrt{\varepsilon_r - \sin^2 \vartheta} \quad . \tag{3.112}$$

Obviously the absolute values of these equations can be written as follows

$$|\,F_\vartheta(\vartheta, \varphi)\,| = g_\vartheta(\vartheta, d_0 + d_1)\, h_\vartheta(\vartheta, d_1)\,\left|\,\frac{F(\vartheta, \varphi)}{F(0,0)}\,\right| \tag{3.113}$$

and

$$|\,F_\varphi(\vartheta, \varphi)\,| = g_\varphi(\vartheta, d_{\hat{0}} + d_1)\, h_\varphi(\vartheta, d_1)\,\left|\,\frac{F(\vartheta, \varphi)}{F(0,0)}\,\right| \quad . \tag{3.114}$$

The inequality (3.108) implies

$$\eta \approx \sqrt{\varepsilon_r} \quad . \tag{3.115}$$

Hence

$$h_\vartheta(\vartheta, d_1) \approx \sqrt{\varepsilon_r} \qquad (3.116)$$

and

$$h_\varphi(\vartheta, d_1) \approx 1 \qquad . \qquad (3.117)$$

Since $F(\vartheta, \varphi)$ can be taken as independent of ϑ and φ according to eqn. (3.101), the relative radiation pattern is a function only of the total thickness of the layer and the angle ϑ.

2) For dipole lengths which are approximately equal to the resonance length or smaller ($l \leq \lambda_e/4$), one can achieve by an appropriate choice of the layer parameters that the magnitude of the electric field strength is the same in the xz- and in the yz-plane. For this, assuming again the validity of the approximation for $F(\vartheta, \varphi)$ (eqn. 3.101)

$$| f_\vartheta(\vartheta) | = | f_\varphi(\vartheta) | \qquad , \qquad (3.118)$$

must hold. We introduce the abbreviations

$$\cos0/1 = \cos(k\sqrt{\varepsilon_{r0/1}} \; d_{\hat{0}/1}) \qquad , \qquad (3.119)$$

$$\sin0/1 = \sin(k\sqrt{\varepsilon_{r0/1}} \; d_{\hat{0}/1}) \qquad (3.120)$$

and the approximations

$$\eta_0 \approx \sqrt{\varepsilon_{r0}} \qquad \text{and} \qquad \eta_1 \approx \sqrt{\varepsilon_{r1}} \qquad . \qquad (3.121)$$

Then we obtain from eqns (3.65) to (3.68)

$$0 = \varepsilon_{r0}\left(\cos0\right)^2 + \varepsilon_{r1}\left(\sin0 \; \text{ctg1}\right)^2 + 2\sqrt{\varepsilon_{r0}\varepsilon_{r1}} \; \cos0 \; \sin0 \; \text{ctg1} \qquad (3.122)$$
$$- \varepsilon_{r0}\varepsilon_{r1}\left(\cos0 \; \cot1\right)^2 - \left(\varepsilon_{r0}\sin0\right)^2 + 2\varepsilon_{r0}\sqrt{\varepsilon_{r0}\varepsilon_{r1}} \; \cos0 \; \sin0 \; \cot1 \quad .$$

The phase difference between observation points in the xz- and in the yz-plane, which have the same value of ϑ, can be calculated from the following equation

$$\Delta \arg = \arg f_\vartheta(\vartheta) - \arg f_\varphi(\vartheta) = \text{arctg}(\frac{F_A}{\cos\vartheta}) - \text{arctg}(F_A \cos\vartheta) \qquad (3.123)$$

with

$$F_A = -\frac{\sqrt{\varepsilon_{r0}} \; \sqrt{\varepsilon_{r1}} \; \cos0 \; \text{ctg1} - \varepsilon_{r0} \sin0}{\sqrt{\varepsilon_{r0}} \; \cos0 + \sqrt{\varepsilon_{r1}} \; \sin0 \; \text{ctg1}} \qquad . \qquad (3.124)$$

If $\varepsilon_{r0} = \varepsilon_{r1} = \varepsilon_r$, eqns. (3.122) and (3.123) can be solved directly. One obtains

$$d_{\hat{0}} + d_1 = \frac{1}{k\sqrt{\varepsilon_r}}\text{arcsin}\sqrt{\frac{\varepsilon_r}{1 + \varepsilon_r}} \qquad (3.125)$$

and

$$\Delta \arg = -\text{arctg}(\frac{1}{\cos\vartheta}) + \text{arctg}(\cos\vartheta) \quad . \tag{3.126}$$

Fig. 3.21 shows the radiation pattern for the current distribution given in Figs. 3.18 and 3.19. The curves a) and c) were drawn for $\varphi = \pi/2$ (yz-plane) as a function of ϑ, the curves b) and d) for $\varphi = 0$ (xz-plane). Eqn. (3.95) implies that in the first case the product

$$f_\varphi(\vartheta) \, F(\vartheta, \varphi = \frac{\pi}{2}) \tag{3.127}$$

is presented and in the second case the product

$$f_\vartheta(\vartheta) \, F(\vartheta, \varphi = 0) \quad . \tag{3.128}$$

We further note that

$$F(\vartheta, \varphi = \tfrac{\pi}{2}) = \text{const} \tag{3.129}$$

for a line current in the x-direction. Therefore, the vertical diagram in the yz-plane is determined only by the Green's function. Altogether, the figures show that by covering a microstrip antenna one has the opportunity to vary the farfield pattern within a wide range.

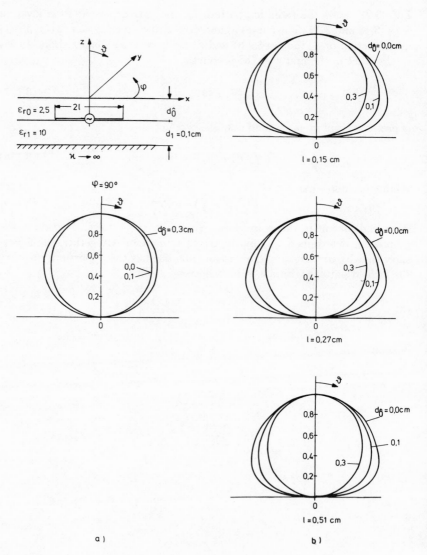

Fig. 3.21 Radiation pattern of printed dipoles

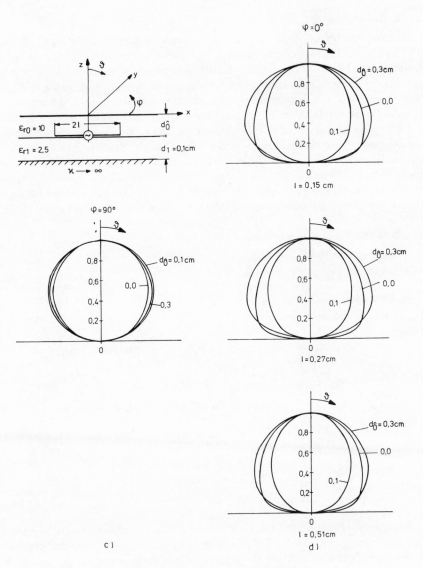

c)   d)

Fig. 3.21 Radiation pattern of printed dipoles

3.4 Arrays of printed dipoles

3.4.1 Basic concept

Printed dipoles are especially suitable for use as elements of planar arrays. The radiating elements are usually fed by a network which is built in microstrip technology as well. The current distribution which is required for a certain radiation pattern can be obtained by using power dividers and phase shifters. The components are usually designed under the assumption that the radiation coupling is negligible. Our discussion of the excitation of surface waves in section 3.2.4.2, however, shows that this assumption has to be checked, especially in those cases where the substrate is not very thin or where a dielectric cover exists. Therefore, we shall study in the next section which currents are induced by a voltage driven dipole in a second dipole. Succeeding sections will show that one can choose parameters in such a way that the currents which have been induced via radiation coupling take on a given amplitude and given phase.

3.4.2 Mutual coupling of two dipole antennas

Consider two thin dipoles of fi-nite length in the plane $z = 0$ (Fig. 3.22). The relationship between currents and voltages at the ports can be described using either an impedance or an admit-tance matrix.

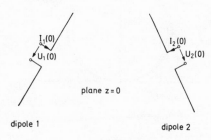

Fig. 3.22 Mutual coupling of two dipoles

$$[U(0)] = [Z][I(0)] \quad , \quad (3.130)$$

$$[I(0)] = [Y][U(0)] \quad . \quad (3.131)$$

The strength of the mutual coupling is given by the off-diagonal elements. Due to reciprocity we have

$$Z_{12} = Z_{21} \tag{3.132}$$

and

$$Y_{21} = Y_{12} \quad . \tag{3.133}$$

For the determination of the matrix elements, we solve the integral equation (2.18) by means of the method of moments. The output of the computer program is the current distribution on the dipoles if the power supply is by a voltage-source at an arbitrary position. Thus one immediately obtains the coupling admittance, if port 1 is connected to a voltage source U_1 and port 2 is short-circuited:

$$Y_{21} = \frac{I_2(0)}{U_1(0)}\Big|_{U_2(0)=0} \quad . \tag{3.134}$$

Large antenna systems often have radiating elements which are arranged in the form of a right angle grid (Fig. 3.23). In order to determine the radiation coupling one therefore first of all has to consider parallel dipoles or dipoles arranged in a line. For large arrays the calculation of the coupling of dipoles far away from each other requires a lot of computer time. In the following, therefore, simple approximate equations will be developed. These are based on the well known exact equation for the coupling of two antennas:

$$Z_{21} = -\frac{\int\limits_{2l_2} \vec{E}_{21}\vec{I}_2 dl}{I_1(0)I_2(0)} \quad . \tag{3.135}$$

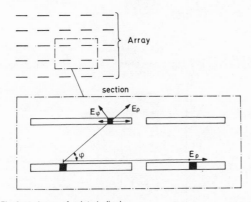

Fig. 3.23 Array of printed dipoles

In our notation $\vec{E}_{21}$ defines the field of dipole 1 at the position of dipole 2 and $\vec{I}_2$ defines the current on dipole 2. As many examples calculated numerically show, the currents are also sinusoidal as a first approximation when excited by radiation coupling, as was the case for the current distribution produced by a voltage source. Thus, eqn. (3.97) is applicable. Furthermore, according to results shown in Fig. 3.15, the surface waves are dominating in the plane $z = 0$. Starting with eqns. (3.44), (3.45) and (2.78) the equations for the surface waves may be approximated in the plane $z = 0$ by

$$dE_{\rho 0 w}^{TM} \approx \frac{\hat{E}_{\rho 0 w}^{TM}}{\sqrt{\rho}}e^{-jk_{\rho\rho}^{TM}\rho}\cos\varphi \, , \quad dE_{\varphi 0 w}^{TM} \approx 0 \quad \text{TM} - \text{surface waves} \tag{3.136}$$

$$dE_{\rho 0 w}^{TE} \approx 0 \, , \qquad dE_{\varphi 0 w}^{TE} \approx \frac{\hat{E}_{\varphi 0 w}^{TE}}{\sqrt{\rho}}e^{-jk_{\rho\rho}^{TE}\rho}\sin\varphi \qquad \text{TE} - \text{surface waves} \tag{3.137}$$

for $\rho \gg \lambda$.

The equations for $\hat{E}_{\rho 0 w}^{TM}$ and $\hat{E}_{\varphi 0 w}^{TE}$ are obtained by comparing with eqns. (3.44) and (3.45). The angle φ (see Fig. 3.24) is approximated by φ_o and the distance ρ by

$$\rho \approx \frac{x - x'}{\cos\varphi_0} \quad . \tag{3.138}$$

We further use

$$\frac{1}{\sqrt{\rho}} \approx \sqrt{\frac{\cos\varphi_0}{x - x'}} \approx \frac{1}{\sqrt{\rho_0}}(1 + \frac{x'}{2\Delta_x}) \quad . \tag{3.139}$$

With these approximations, the integrals are

$$\vec{E}_{21}(x) = \vec{u}_\rho \hat{E}_{\rho 0 w}^{TM} I_1(0) \int_{-l_1}^{+l_1} \frac{e^{-j(x-x')\frac{k_{\rho p}^{TM}}{\cos\varphi_0}}}{\sqrt{\rho_0}}(1 + \frac{x'}{2\Delta_x})\cos\varphi_0 \cos(k_e x')\,dx'$$

$$\tag{3.140}$$

$$+ \vec{u}_\varphi \hat{E}_{\varphi 0 w}^{TE} I_1(0) \int_{-l_1}^{+l_1} \frac{e^{-j(x-x')\frac{k_{\rho p}^{TE}}{\cos\varphi_0}}}{\sqrt{\rho_0}}(1 + \frac{x'}{2\Delta_x})\sin\varphi_0 \cos(k_e x')\,dx' \, .$$

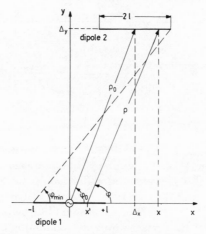

Fig. 3.24 Arrangement of staggered dipoles

Introducing the abbreviations

$$\frac{k_{\rho p}^{TM}}{\cos\varphi_0} = k_\rho^M \quad \text{and} \quad \frac{k_{\rho p}^{TE}}{\cos\varphi_0} = k_\rho^E \quad , \tag{3.141}$$

the integrals are solved analytically for two $\lambda_e/2$-dipoles:

$$\vec{E}_{21}(x) \approx \vec{u}_\rho \hat{E}_{\rho 0 w}^{TM} I_1(0)\cos\varphi_0 \, \frac{e^{-jk_\rho^M x}}{\sqrt{\rho_0}}L^M + \vec{u}_\varphi \hat{E}_{\varphi 0 w}^{TE} I_1(0)\sin\varphi_0 \, \frac{e^{-jk_\rho^E x}}{\sqrt{\rho_0}}L^E$$

$$\tag{3.142}$$

with

$$L^M = L_1(k_\rho^M) + \frac{1}{2\Delta_x} L_2(k_\rho^M) \quad , \tag{3.143}$$

$$L^E = L_1(k_\rho^E) + \frac{1}{2\Delta_x} L_2(k_\rho^E) \quad , \tag{3.144}$$

$$L_1(\xi) = \frac{2k_e}{k_e^2 - \xi^2} \cos(\xi \frac{\lambda_e}{4}) \quad , \tag{3.145}$$

$$L_2(\xi) = \frac{j\pi}{k_e^2 - \xi^2} \sin(\xi \frac{\lambda_e}{4}) - \frac{4jk_e\xi}{(k_e^2 - \xi^2)^2} \cos(\xi \frac{\lambda_e}{4}) \quad . \tag{3.146}$$

Subsequently the integration over the current distribution of dipole 2 is performed resulting in

$$
\begin{aligned}
Z_{21} \approx & -\frac{\hat{E}_{\rho 0w}^{TM}}{\sqrt{\rho_0}} \cos^2\varphi_0 \, e^{-jk_\rho^M \Delta_x} L_1(k_\rho^M) \left(L_1(k_\rho^M) + \frac{1}{2\Delta_x} L_2(k_\rho^M) \right) \\
& + \frac{\hat{E}_{\varphi 0w}^{TE}}{\sqrt{\rho_0}} \sin^2\varphi_0 \, e^{-jk_\rho^E \Delta_x} L_1(k_\rho^E) \left(L_1(k_\rho^E) + \frac{1}{2\Delta_x} L_2(k_\rho^E) \right) \quad . \tag{3.147}
\end{aligned}
$$

Eqn. (3.147) is not applicable for $\varphi_0 = \pi/2$ because of the terms

$$k_\rho^{M/E} = \frac{k_{\rho\rho}^{TM/TE}}{\cos\varphi_0} \quad . \tag{3.148}$$

This case therefore must be treated separately. According to Fig. 3.24 we have:

$$\rho = \sqrt{(x - x')^2 + \Delta_y^2} \approx \Delta_y + \frac{(x - x')^2}{2\Delta_y} \quad , \tag{3.149}$$

$$\frac{1}{\sqrt{\rho}} \approx \frac{1}{\sqrt{\Delta_y}} - \frac{(x - x')^2}{4\Delta_y^2 \sqrt{\Delta_y}} \quad , \tag{3.150}$$

$$\varphi_{min} = \arctan(\frac{2\Delta_y}{\lambda_e}) \approx \pi/2 \quad , \tag{3.151}$$

$$\cos\varphi \approx 0 \quad , \qquad \sin\varphi \approx 1 \quad . \tag{3.152}$$

With the further approximation, that

$$e^{-jk_{\rho\rho}^{TE}\frac{(x - x')^2}{2\Delta_y}} \approx 1 \tag{3.153}$$

the integral for $\vec{E}_{21}$ can be solved:

$$\vec{E}_{21}(x) \approx \vec{u}_\varphi \hat{E}_{\varphi 0w}^{TE} I_1(0) \frac{e^{-jk_{\rho\rho}^{TE}\Delta_y}}{\sqrt{\Delta_y}} \left(\frac{\lambda_e}{\pi} - \frac{\lambda_e}{4\pi y_0^2} \left(x^2 + (\frac{\lambda_e}{4})^2 - (\frac{\lambda_e}{\pi})^2 \right) \right) \quad . \tag{3.154}$$

The result for the mutual coupling of two parallel $\lambda_e/2$-dipoles is

$$Z_{21} \approx \hat{E}_{\varphi 0 w}^{TE} \frac{4}{k_e^2} \frac{e^{-jk_{\rho\rho}^{TE}\Delta_y}}{\sqrt{\Delta_y}} \left(1 - (\frac{1}{2} - \frac{4}{\pi^2})(\frac{\lambda_e}{4\Delta_y})^2 \right) \quad . \tag{3.155}$$

Figs. 3.25a to 3.25d show a few examples of coupling admittances calculated numerically compared with the results given by the approximate eqns. (3.147) and (3.155). The dipole lengths have been determined according to equation

$$l = \frac{1}{4} \frac{\lambda}{\sqrt{\frac{\varepsilon_{r0}+\varepsilon_{r1}}{2}}} \tag{3.156}$$

so that the dipoles roughly operate in resonance. A common feature of all figures is that G_{12} and B_{12} for $\Delta_{x,y} > 2$ cm are oscillatory with decreasing amplitude. This is easily explained with the help of eqn. (3.135). According to this equation $\vec{E}_{21}$ and therefore Z_{21} as well are approximately proportional to the Green's function with

$$|\vec{r}' - \vec{r}| \approx \Delta_{x,y} = \sqrt{\Delta_x^2 + \Delta_y^2} \tag{3.157}$$

in the case of two short dipoles at a large distance from each other. This special case does not occur among the examples shown here but the results approximately confirm this statement. Thus, only the TM_1-surface wave can propagate on the structure the results shown in Figs. 3.25a and 3.25b are based on. Since for large distances

$$|dE_{\varphi w}^{TM_1}| \ll |dE_{\rho w}^{TM_1}| \quad , \tag{3.158}$$

the radiation coupling of dipoles arranged in parallel is smaller than that of dipoles arranged in a line. In the latter case the peaks of G_{12} and B_{12} in the range $\Delta_{x,y} > \lambda$ decay roughly proportionally to $\sqrt{\Delta_x}$ (Fig. 3.25a). The distance between two zeros of G_{12} and B_{12} is approximately

$$\frac{\lambda^{TM_1}}{2} = \frac{\pi}{k_{\rho\rho}^{TM_1}} \quad . \tag{3.159}$$

The TM_1- and TE_1-surface waves can propagate on the structures shown in Figs. 3.25c and 3.25d. Since coupling of dipoles in a line only occurs via the E_ρ-components and thus via the TM_1-wave, the results resemble those of Fig. 3.25a. The curves in Fig. 3.25d for dipoles arranged parallel to each other, however, differ greatly from those in Fig. 3.25b. This is due to the fact that the field strength $\vec{E}_{21}$ is equal to the superposition of $E_\varphi^{TE_1}$ and $E_\rho^{TM_1}$, the amplitudes of both parts being roughly equal and the phases being very different.

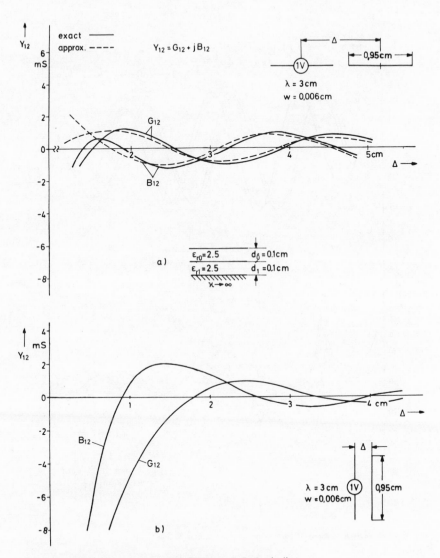

Fig. 3.25 Mutual coupling of parallel dipoles and dipoles in line

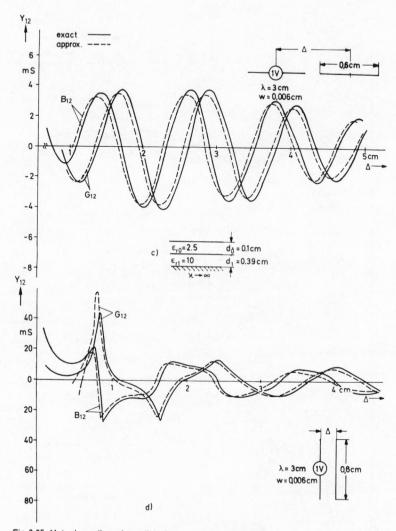

Fig. 3.25 Mutual coupling of parallel dipoles and dipoles in line

If one calculates radiation coupling of dipoles in or on those structures that facilitate the excitation of several surface waves one gets results that are virtually impossible to interpret.

In Fig. 3.25 we have also plotted the results which have been obtained with the help of the approximate eqns. (3.147) and (3.155). In all cases which have been investigated the approximations yield smaller amplitudes than the exact solution. Furthermore the phases deviate significantly. Although it is relatively simple to determine correction factors, this is not necessary because the approximate coupling impedances are only used for dipoles far away from each other.

3.4.3 Arrays fed by voltage sources

The main problem with large arrays is the limited storage capacity of the computer because each dipole must be divided into several subsegments when applying the method of moments. In order to get reliable results for the input impedance, about 10 segments per wavelength are required for a single dipole (diameter 0.01 λ). The number of segments which should be used for the calculation of the mutual coupling of two dipoles depends on the distance between the dipoles. According to experience gained so far, at least 5 segments are necessary for two half-wavelength dipoles in line, if the distance between the adjacent ends is about $\lambda/4$. As the order of the linear system of equations is equal to $N \cdot M \cdot q$ (N, M are the numbers of dipoles in a row and a column respectively (Fig. 3.26), q is the number of expansion modes, which is equal to the number of segments minus one), obviously, only arrays consisting of some few dipoles can be treated in this way.

The search for an alternative starts with results obtained by numerical experiments. First, the self-impedance Z_{AA} was calculated for the dipoles in the centre of arrays with a) 3×3, b) 5×5 and c) 7×7 dipoles (dipole A in Fig. 3.27). Each dipole was divided into 8 segments corresponding to $q = 7$. The difference between Z_{AA} of cases b) and c) was typically less than 2%, for a) and b) up to 10%. This shows that dipoles far away from each other do not significantly affect the self-impedance.

In a next step (case d)), a 7×7 array was investigated with $q = 7$ for the centre dipole A, $q = 5$ for the adjacent dipoles (rings B and C in Fig. 3.27) and $q = 3$ for the dipoles at the rim (ring D). The difference between the results of c) and d) is less than 2% for Z_{AA} and up to 5% for the mutual coupling coefficients of dipoles located in ring B. This again stresses the fact that only a limited number of dipoles close to the element under consideration will significantly influence the self- and mutual impedances. Therefore, the coefficients of the system of linear equations for large arrays were determined by the following procedure: As a first step, an array with 7×7 dipoles was investigated in order to determine the self-impedance of the centre dipole Z_{AA} and the mutual impedances of dipoles belonging to the rings B and C. Then, the diagonal elements of the system

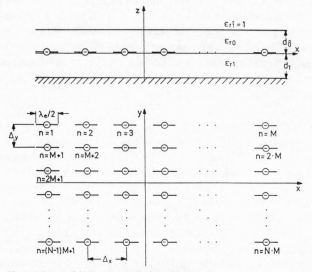

Fig.3.26 Array of N·M printed dipoles with a cover

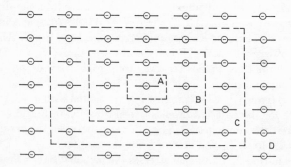

Fig.3.27 Array of 7 x 7 dipoles

of $N \cdot M$ linear equations were set equal to Z_{AA}, the non-diagonal elements equal to the mutual impedances as far as available. The remaining non-diagonal elements, which belong to elements far away from each other, were calculated using only one expansion mode ($q = 1$). The elements belonging to dipoles located at the rim of the array were treated in a similar way. Thus, a system of $N \cdot M$ linear equations was obtained and solved in order to determine the input currents of the dipoles.

Of course, the described procedure is very tedious. The main reason is that the Green's function consisting of integrals of Sommerfeld type must be determined up to a distance between observation and source points of

$$\rho_{max} = \sqrt{(M \, \Delta_x)^2 + (N \, \Delta_y)^2} \quad . \tag{3.160}$$

For an array of, for example, 15×15 dipoles and $\Delta_x = \Delta_y = 0.5\lambda$ we have $\rho_{max} \approx 10.6\lambda$. Therefore, for large distances the approximate eqns. (3.147) and (3.155) which have been derived previously can be applied.

In the following we shall discuss the current distributions on some arrays with dielectric covers. As a first example, Figs. 3.28a and 3.28b show an array of 5×5 dipoles. Because of the symmetry only one quarter is illustrated. The distance between two adjacent elements is $0.5 \, \lambda_0$, the dielectric constants of the substrate and the cover are 2.5. With these parameters, only a TM_1-surface wave with a small amplitude can propagate. Consequently, the mutual coupling is small, as well. For the results in Fig. 3.28a (normalized amplitudes) and Fig. 3.28b (phases) all 25 elements have been supplied in phase by a voltage source of 1 V in order to get broadside radiation. The minimum amplitude $|I_{min}|$ occurs on the dipoles $n = 2, 4, 22, 24$, the maximum amplitude $|I_{max}|$ on the dipoles n $= 8, 18$ with $|I_{min}|/|I_{max}| = 0.43$. The maximum deviation of the phases from zero is 15°. Probably it will be more useful to use the standard deviation from the arithmetic means:

$$\sigma = \sqrt{\frac{1}{n-1} \sum_{i=1}^{n} (a_i - a)^2} \quad . \tag{3.161}$$

In the example discussed above $\sigma_{ph} \approx 7°$ is obtained for the phases and $\sigma_{am} \approx 0.21|I_{max}|$ for the amplitudes.

As an example for a large array, Fig. 3.28c shows the results for 17×17 dipoles (again only one quarter is displayed). The current distribution of the large array is smoother than that of a small one. This is easily understood noting that an infinite array must have the same amplitudes and phases. The standard deviations are now $\sigma_{ph} \approx 6°$ and $\sigma_{am} \approx 0.09|I_{max}|$.

Fig. 3.29 illustrates the currents of a 5×5 element array, the main beam of which is shifted away from the broadside direction. The dashed lines mark the phases of the input voltages, the full lines the phases of the input currents.

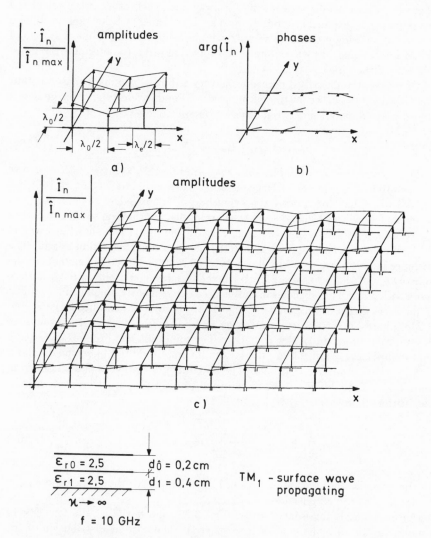

Fig. 3.28 Normalized input currents of an array of 5x5 (a/b)
and of 17x17 (c) dipoles. Phases of the input voltages 0°
(Because of symmetry one quarter only represented)

Because of the asymmetric voltage supply the current distribution is asymmetric too.

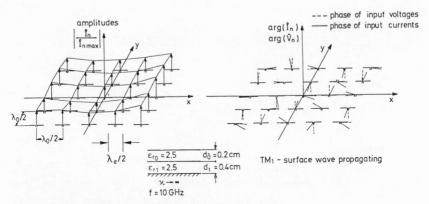

Fig.3.29 Normalized input currents for an array of 5x5 dipoles.
Phases of the input voltages n·90°

The standard deviation for the phases is $\sigma_{ph} \approx 15°$. The amplitude distribution is rather smooth ($\sigma_{am} \approx 0.11|I_{max}|$) with slightly increasing amplitudes towards the left fringe.

The next two examples are discussed for a structure with a strong TE_1-surface wave and no TM_1-surface wave excitation (this structure will be discussed in more detail in the next section). For the results in Fig. 3.30a and 3.30b (5×5 dipoles) the voltage sources have again been taken in phase. The corresponding current distribution is rather smooth ($\sigma_{am} \approx 0.13|I_{max}|$), but the value for σ_{ph} comes up to 24° because of the strong coupling by the TE_1- surface wave. This fact is emphasized by the results obtained for an array of 17×17 dipoles (Fig. 3.30c). The dipoles in line (x-direction) show a smooth amplitude distribution, while parallel to the dipoles (y-direction) there exist significant oscillations. The values for the standard deviations are now $\sigma_{ph} \approx 42°$ and $\sigma_{am} \approx 0.19|I_{max}|$.

If the array is scanned, the surface wave coupling can lead to scan blindness [3.113], where only a reduced part of the power is radiated. True scan blindness with a reflection coefficient equal to unity occurs only for infinite arrays. For this case Pozar and Schaubert [3.83] have derived the simple equation

$$(k_{\rho\rho}^{TM/TE})^2 = \left(\frac{p\lambda}{a} + \sin\vartheta_{bl} \cos\varphi_{bl} \right)^2 + \left(\frac{q\lambda}{b} + \sin\vartheta_{bl} \sin\varphi_{bl} \right)^2 \qquad (3.162)$$

for the blind angle ($\vartheta_{bl}, \varphi_{bl}$) starting with the plane wave spectrum representation instead of the cylindrical wave spectrum used within this book. In this

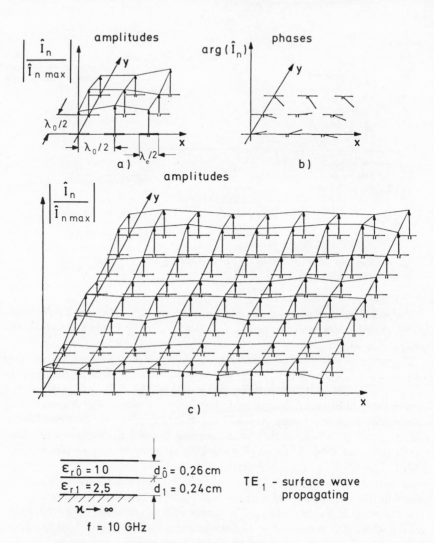

Fig. 3.30 Normalized input currents of an array of 5x5 (a/b)
          and of 17x17 (c) dipoles. Phases of the input voltages 0°
          (Because of symmetry one quarter only represented)

equation, p, q are integers in the range $-\infty < p, q < +\infty$. Only those numbers p, q must be used which lead to blind angles $(\vartheta_{bl}, \varphi_{bl})$ within the visible space. For the discussion of the results, Pozar defined a reflection coefficient $R_n(\vartheta, \varphi)$ of the n'th element by the equation

$$R_n(\vartheta, \varphi) = \frac{Z_{in}^n(\vartheta, \varphi) - Z_{in}^n(0,0)}{Z_{in}^n(\vartheta, \varphi) + Z_{in}^n(0,0)^*} \tag{3.163}$$

for arrays supplied with currents of uniform amplitudes and scanned in the (ϑ, φ) direction. Whereas this reflection coefficient is very useful for arrays of infinite size (in this case the input impedances of all elements are equal) or very large size, its application is less convenient for small arrays, because the results strongly depend on the element position. This problem is overcome by use of the arithmetic mean of the real part of the input impedance $\bar{R}_{in}$.

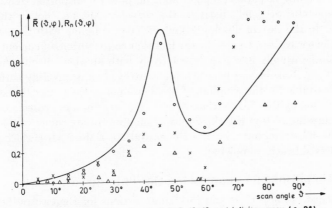

Fig. 3.31 Reflection coefficient of a 5x5, 7x7, 19x19 and infinite array ($\varphi = 0°$)
——— infinite [3. 113]
o o o o 19x19
x x x x 7x7
△ △ △ △ 5x5

Because of

$$P_{in} = \frac{1}{2} \sum_{n=1}^{N \cdot M} Re\{Z_{in}^n\} |I_n(0)|^2 \tag{3.164}$$

this arithmetic mean $\bar{R}_{in}$ is equal to the power P_{in} of the array divided by the number of elements, if all dipoles are fed with the input currents $|I_n(0)| = |I(0)| = 1A$, $n = 1, ..., N \cdot M$. Then the equation for the mean value of the reflection coefficient $\bar{R}(\vartheta, \varphi)$ reads

$$\bar{R}(\vartheta, \varphi) = \frac{\bar{R}_{in}(\vartheta, \varphi) - \bar{R}_{in}(0,0)}{\bar{R}_{in}(\vartheta, \varphi) + \bar{R}_{in}(0,0)} \quad . \tag{3.165}$$

In Fig. 3.31 the results are plotted for arrays 5×5, 7×7, 19×19 and an infinite array. It shows that blind angles occur even for small arrays. As was to be expected, the reflection coefficients of a 19×19 array resemble those of an infinite array.

3.4.4 Arrays fed by radiation coupling

3.4.4.1 Some basic considerations

Arrays in which part of the radiating elements is fed by radiation coupling are very important in antenna technology. The Yagi antenna should be mentioned as the probably most important application of this principle. If a Yagi antenna is constructed using usual microstrip technology as is suggested for example in [3.18] and [3.19], then the following should be taken into account: since, in the case of a thin substrate, only the TM_1-surface wave is excited, there is only minimal coupling of parallel dipoles and the power transported by the surface wave propagates perpendicularly to the antenna axis, i.e. not in the desired direction. In the case of dipoles in line, effective coupling with the help of the TM_1-surface wave is prevented by the fact that contributions of current elements on one dipole usually interfere destructively with another. If one chooses the parameters of the substrate in such a way as to excite a strong TE_1-surface wave, which is suitable for the coupling of parallel dipoles, the power transported at the same time by the TM_1-surface wave is too large in most cases.

The following section will show that by an appropriate choice of the parameters of the substrate and the cover the excitation of the TM_1-surface wave can be suppressed nearly completely.

3.4.4.2 Conditions for suppressing the TM- surface waves

With eqns. (2.56), (3.9) and (3.11) one obtains for the electric field of the TM-surface waves in layer "1", i.e. the substrate:

$$dE_{\rho 1 w}^{TM} = -2\pi j M \frac{-jB_{0p}}{k_{\rho p}^{TM}} \sin(k_{z1p}^{TM}(d_1 + z)) \, H_1^{(2)'}(k_{\rho p}^{TM}\rho)\cos\varphi \quad , (3.166)$$

$$dE_{\varphi 1 w}^{TM} = -2\pi j M \frac{jB_{0p}}{(k_{\rho p}^{TM})^2 \rho} \sin(k_{z1p}^{TM}(d_1 + z)) \, H_1^{(2)}(k_{\rho p}^{TM}\rho)\sin\varphi \quad , (3.167)$$

$$dE_{z1 w}^{TM} = -2\pi j M \frac{jB_{0p}}{k_{z1p}^{TM}} \cos(k_{z1}^{TM}(d_1 + z)) \, H_1^{(2)}(k_{\rho p}^{TM}\rho)\cos\varphi \quad , \quad (3.168)$$

with B_{0p} from eqn. (3.50) and

$$M = \frac{1}{\sin(k_{z1p}^{TM}d_1)} \left(k_{zp}^{TM} \cos(k_{z0p}^{TM}d_{\hat{0}}) + j\frac{k_{z0p}^{TM}}{\varepsilon_{r0}} \sin(k_{z0p}^{TM}d_{\hat{0}}) \right) e^{-jk_{zp}^{TM}d_{\hat{0}}} \quad .$$

$$(3.169)$$

In order to determine $k_{\rho p}^{TM}$ eqn. (3.27) must be solved. The electric field of TM-surface waves approaches zero in the layer "1" if $M = 0$. Therefore, in addition to eqn. (3.27) the equation

$$k_{zp}^{TM} \cos(k_{z0p}^{TM} d_{\hat{0}}) + j \frac{k_{z0p}^{TM}}{\varepsilon_{r0}} \sin(k_{z0p}^{TM} d_{\hat{0}}) = 0 \qquad (3.170)$$

must also be satisfied (in this section it is again assumed that $\varepsilon_{ri}'' = 0$). Eqn. (3.170) is identical with eqn. (3.27) for

$$k_{\rho p}^{TM} = k_1 = k\sqrt{\varepsilon_{r1}} \qquad . \qquad (3.171)$$

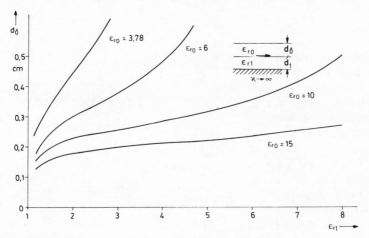

Fig.3.32   Values for $\varepsilon_{r\hat{0}} . \varepsilon_{r1}$ and $d_{\hat{0}}$ for suppressing the TM_1 -surface wave

In the case of a surface wave the field has to decay exponentially in the space above the layers - as has already been discussed in section 3.2.3. Therefore,

$$k_{zp}'' = Im\left\{ \sqrt{k^2 - k_{\rho p}^2} \right\} \qquad (3.172)$$

has to be less than zero. This is only fulfilled, according to eqn. (3.4), if k_{z0p} is real. It follows that because of

$$k_{z0p} = \sqrt{k_0^2 - k_{\rho p}^2} = \sqrt{k_0^2 - k_1^2} \qquad (3.173)$$

the dielectric constant of the cover must be higher than that of the substrate, hence

$$\varepsilon_{r0} > \varepsilon_{r1} \qquad . \qquad (3.174)$$

Hence, from eqn. (3.170)

$$d_{\hat{0}} = \frac{1}{k\sqrt{\varepsilon_{r0} - \varepsilon_{r1}}} \left(\text{arctg}(\frac{\sqrt{\varepsilon_{r1} - 1}}{\sqrt{\varepsilon_{r0} - \varepsilon_{r1}}} \varepsilon_{r0}) + n_d \pi \right) ; \quad n_d = 0, \pm 1, \pm 2, \cdots.$$

(3.175)

In Fig. 3.32 $d_{\hat{0}}$ is plotted as a function of ε_{r1} for some values of $\varepsilon_{r\hat{0}}$ and $n_d = 0$. Note that eqn. (3.175) is independent of d_1, hence the value for d_1 can be chosen according to other criteria. If one chooses

$$\varepsilon_{r0} = 10 \qquad \text{and} \qquad \varepsilon_{r1} = 2.5 \quad , \tag{3.176}$$

according to the previous examples, eqn. (3.175) yields

$$d_{\hat{0}} = 0.24 \text{ cm} \quad . \tag{3.177}$$

The thickness of the substrate can be established according to various criteria. In order to achieve a strong coupling, first of all d_1 will be chosen in such a way that the TE_1-surface wave is strongly excited. To this end we have plotted the absolute values of the field strengths for $d_{\hat{0}} = 0.24$ cm as a function of d_1 in Fig. 3.33. We see that d_1 can be in the range

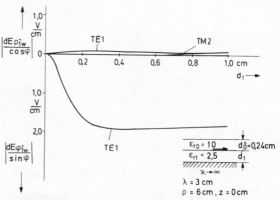

Fig.3.33 Amplitudes of the TE_1 - and TM_2 surface waves excited by a point source $I_x dl_x \sqrt{\mu_0 / \varepsilon_0} / 4\pi = 1$ Vcm)

$$0.26 \text{ cm} \lesssim d_1 < 0.65 \text{ cm}. \tag{3.178}$$

Further, one can see from Fig. 3.33 that the TM_1-surface wave is not excited. For $d_1 = 0.26$ cm the coupling admittance of parallel dipoles and dipoles in line is plotted in Fig. 3.34. It is substantially smaller in the latter case than in the former. If the distance, for example, is $\Delta = \lambda/2 = 1.5$ cm, then

$$| Y_{12} | = 14.5 \text{ mS} \qquad \text{(parallel dipoles)} \tag{3.179}$$

$$| Y_{12} | = 2.5 \text{ mS} \qquad \text{(dipoles in line)} . \tag{3.180}$$

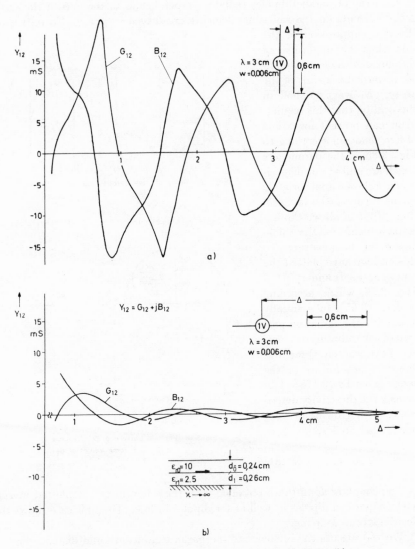

Fig. 3.34 Mutual coupling of two parallel dipoles (a) and two dipoles in line (b)
(TM₁ - surface wave suppressed)

3.4.4.3 Array of radiation-fed parallel dipoles for broadside radiation

An array of parallel dipoles radiates perpendicular to the axis of the array if the currents on the radiating elements are driven in phase. To obtain an effective suppression of the side lobes the amplitudes of the currents have to diminish towards both ends of the array. We want to show in this section that this requirement can be met for a line of five elements according to Fig. 3.35, if the parameters which are still at our disposition are chosen appropriately in the setup of the last section. First of all we chose a suitable value for the thickness d_1 of the substrate. To this end we have plotted the functions $f_\vartheta(\vartheta)$ and $f_\varphi(\vartheta)$ in Fig. 3.36, which, according to eqn. (3.95), substantially determine the radiation pattern of one radiating element. We have chosen three different values for d_1 in the range given by (3.178). Obviously the directivity deteriorates with growing d_1. For our further analysis we chose

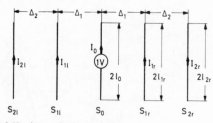

Fig. 3.35 Array of five parallel dipoles

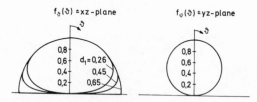

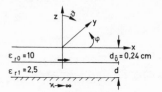

Fig. 3.36 Farfield pattern of point source $I_x dl_x$ for some values of d_1

$$d_1 = 0.26 \text{ cm} \quad . \quad (3.181)$$

The coupling admittance obtained with these parameters is plotted in Fig. 3.34 for parallel dipoles as well as for dipoles in line. The first case shows the desired strong coupling.

We will discuss the influence of the parameters length and distance first of all qualitatively. Based on the results presented in section 3.4.2 it follows that the phase of $\vec{E}_{21}$, i.e. the phase of the electric field produced by dipole "1" at the position of another dipole "2", is effectively determined by the propagation constant of the TE_1-surface wave and the distance Δ, i.e.

$$\arg(\vec{E}_{21}) = -k_{\rho\rho}^{TE_1}\Delta + C_q \quad . \quad (3.182)$$

C_q is a constant which is of no interest here. If one neglects the mutual coupling, the short-circuit current $I_2(0)$ of dipole "2" is proportional to the open-circuit voltage produced by field $\vec{E}_{21}$ and the admittance Y_{22}. The value of $\arg(Y_{22})$ can be varied over a wide range by varying the dipole length, as can be seen from the results in section 3.3.1.

In the present case the mutual coupling certainly cannot be neglected because of the short distance ($\Delta_{x,y} < \lambda/2$). Therefore, the currents produced by radiation coupling are determined numerically. As a first step the parameters of the two dipoles S_{1l} and S_{1r} on the left and right side of the centre dipole (S_0) are determined. Dipole S_0 is driven by a voltage source. The amplitudes and phases of the currents $I_{1l/r}$ on these dipoles as a function of the distance Δ_1 are given in Fig. 3.37a for three different lengths of $S_{1l/r}$. The length of S_0 was kept constant at

$$2l_0 = \frac{\lambda}{2} \frac{1}{\sqrt{\frac{\varepsilon_{r0} + \varepsilon_{r1}}{2}}} = 6 \text{ mm} \quad . \tag{3.183}$$

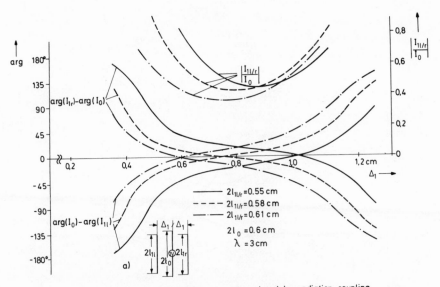

Fig. 3.37 Amplitudes and phases of the dipole currents induced by radiation coupling

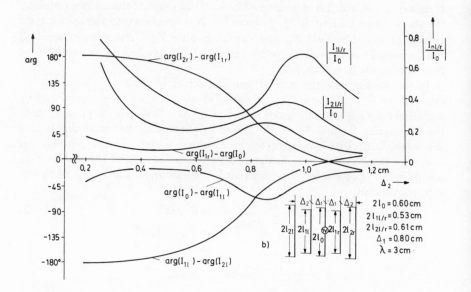

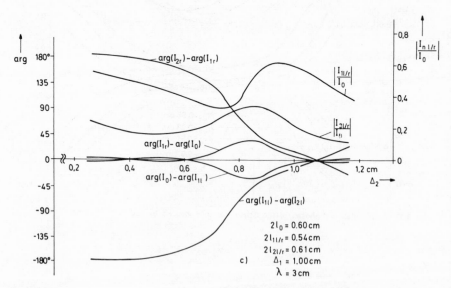

Fig. 3.37 Amplitudes and phases of the dipole currents induced by radiation coupling

One can see that in each of the examples shown, there is a value for Δ_1 for which the phases of I_0 and $I_{1l/r}$ are identical. The amplitudes of $I_{1l/r}(0)$ relative to $I_0(0)$ are in all cases approximately equal for this distance.

As a second step the currents are calculated with two more dipoles $S_{2l/r}$ (Figs. 3.37b and 3.37c). Obviously the currents on $S_{2l/r}$ influence those on $S_{1l/r}$, but this influence, however, turns out not to be very strong. Therefore, for the new calculations the values found before for Δ_1 and $I_{1l/r}$ are chosen as starting points and both parameters are only varied in the range

$$5.2 \text{ mm} \leq 2l_{1l/r} \leq 5.8 \text{ mm} \quad , \tag{3.184}$$

$$7.0 \text{ mm} \leq \Delta_1 \leq 12.0 \text{ mm} \quad . \tag{3.185}$$

Just a few calculations show that $l_{2l/r}$ must be chosen as follows

$$l_{2\,l/r} > l_{1\,l/r} \tag{3.186}$$

in order to get equal phases of the currents for a value of Δ_2 in the range between $\lambda/4$ and $\lambda/2$. Figs. 3.37a and 3.37b present two examples of many cases calculated. From the latter example we deduce the following practicable design prescriptions

$$2l_0 = 6.0 \text{ mm}, \quad 2l_{1\,l/r} = 5.4 \text{ mm}, \quad 2l_{2\,l/r} = 6.1 \text{ mm} \tag{3.187}$$

$$\Delta_1 = 10 \text{ mm}, \quad \Delta_2 = 11 \text{ mm} \quad . \tag{3.188}$$

The phases of the currents on the two extreme dipoles then differ by $7°$, and the normalized amplitudes read

$$\mid I_0 \mid = 1.000 \quad , \tag{3.189}$$

$$\mid I_{1\,l/r} \mid = 0.540 , \tag{3.190}$$

$$\mid I_{2\,l/r} \mid = 0.160 . \tag{3.191}$$

With the known antenna currents the radiation pattern can be determined by applying eqn. (3.95). Fig. 3.38 shows the radiation pattern of the array. In the xz-plane the pattern is equal to that of one dipole in the structure. Because of the short dipoles ($2l \approx 0.2\lambda_0$) this pattern is only determined by the parameters of the layers. In the yz-plane the pattern is given by the product of the element pattern and the array factor. The latter has no sidelobes as a result of the decreasing currents $|I_2| < |I_1| < |I_0|$.

As a last step it was proved that the main part of the power carried by the TE_1-surface wave is radiated by the dipoles, otherwise the antenna would be useless. Therefore, the electric field was calculated on the y-axis in order to determine the amplitude of the TE_1-surface wave (Fig. 3.39). Curve a) shows the field radiated by one dipole, curve b) by 3 and curve c) by 5 dipoles for equal input power. The amplitude of the TE_1-surface wave of case b) is 0.22 times the amplitude of case a), which means that 95% of the power carried by the TE_1-surface wave is radiated. In case c) the level of the TE_1-surface wave is higher than in case b), because the parameters of the array have been chosen for equal phase of the antenna currents and not for maximum radiated power.

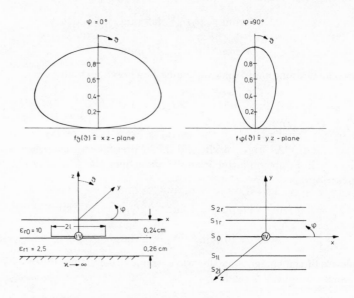

Fig. 3.38 Radiation pattern of five parallel dipoles fed by radiation coupling

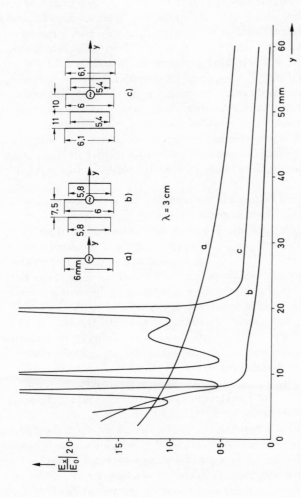

Fig. 3.39 Normalized electric field of one dipole (a) and an array of three dipoles (b) and five dipoles (c) transverse to the dipole (s) for equal input power

3.5 Summary of Chapter 3

Printed antennas have been used frequently since the seventies because of their favourable properties, e.g. low weight, flat construction and therefore good aerodynamic properties, low cost production and good integrability. Up to about 1980 the various methods for the analysis of their radiation properties were based on rather simple approximations, which, however, yield good results as long as the substrate used is electrically thin. An exact formulation of the integral equation for the antenna current is possible with the help of the concept for stratified media, which is used in this book. In section 3.2.1 the equations for the E_z^{TM}– and the H_z^{TE}–component for Green's function of microstrip structures with and without dielectric covering are given. As the integrands can be written down explicitly (and not only with the help of recursive formulas), some analytical properties of the integrand can be discussed in more detail, as in section 2.4.1. The asymptotic interpretation of the integrals with the help of the saddle-point method (section 3.2.3) shows that apart from the space wave, which is represented by the saddle–point contribution, surface and leaky waves can appear.

Special care should be taken in the investigation of the propagation of surface waves (section 3.2.4). Their propagation constants result from the determination of the poles of the integrands. For thick substrates and/or thick superstrates higher surface waves will propagate apart from the TM_1-surface wave, which is always able to exist. If ε_r of the superstrates is smaller than that of the substrate, the dependence of the propagation constants on the layer thickness is roughly equal to that of only one layer. Qualitatively different results, however, are obtained in the reverse case. The amplitudes of the surface waves which are excited by a point source can be calculated with the help of the residue theorem. The results prove that surface waves are negligible for structures with thin substrates. This is why the results which have been obtained with the help of approximation procedures are so good in these cases. On the other hand, if one uses thick substrate layers (e.g. to improve the bandwidth), or if one puts a dielectric covering onto the structure, then surface waves with a very large amplitude will be excited under certain circumstances.

For the propagation constant as well as for the amplitude of the TM_1- surface waves, simple approximation equations are derived for the case of thin layers.

The equations for the space wave of a point source can be derived very easily from the general integral representation (section 3.2.5). Their evaluation shows that by a suitable choice of the four parameters at hand — thicknesses and dielectric constants of substrate and superstrate — a multitude of different diagram shapes can be generated.

Solutions for observation points up to the distance of a few wavelengths are obtained in section 3.2.6 with the help of numerical integration. The field singularity for $\rho \to 0$ and also the asymptotic behaviour are distinctly visible. The latter is in some cases valid already for $\rho \gtrsim \lambda/2$ in contrast to the assumption

$\rho \gg \lambda$ needed in the derivation. By comparison of the numerical results with those which have been calculated using the approximation equation for very small distances, one recognizes the very small region of validity of these equations ($0 < \rho \lesssim \lambda/100$). Still, they prove to be very useful for the calculation of self–impedances.

In section 3.3 dipoles of finite length are investigated. First of all, in section 3.3.1 the current distribution and the input impedance are calculated. It turns out that the relative current distribution is mainly determined by the length of the dipoles and by the material in close proximity to the dipoles. The input impedance on the other hand is influenced by further parameters, e.g. the thickness of substrate and superstrate.

The radiation field of dipole antennas of finite length is calculated from the current distribution by integration and multiplication of Green's function (section 3.3.2). Some results should be specially emphasized:

1) For $\varepsilon_{re} \gg 1$ the radiation pattern of a $\lambda_e/2$ – resonator is nearly exclusively determined by Green's function of the structure and hardly at all by the current distribution on the dipole.
2) For dipole antennas on a thin substrate the power radiated by the space wave grows proportional to d_1^2, the power taken by the surface wave grows proportional to d_1^3.
3) For $\varepsilon_{ro/1} \gg 1$ the following holds:
   a) If substrate and superstrate are made of the same material, the radiation diagram will be determined just by the thickness of the total dielectric layer and not by the thicknesses of the single layers.
   b) If the dipole length is about equal to or smaller than the first resonance length, the same shape of the pattern can be achieved for the E– and the H–plane by an appropriate choice of parameters.

Section 3.4 is concerned with the radiation behaviour of arrays of printed dipoles. After some fundamental considerations (section 3.4.1), first of all the radiation coupling of two dipoles only is investigated. As the radiation coupling is mainly effected by the surface waves which are able to propagate, approximation equations can be derived. They reveal that dipoles in parallel are coupled mainly by TE-, dipoles in line mainly by TM- surface waves. By comparison with numerical results the accuracy of the approximation equation is assessed. Section 3.4.3 deals with arrays of finite size, where every antenna is fed by an impressed source. The analysis of large arrays leads to very large systems of equations. Therefore, a procedure is developed that can reduce the computational effort for large arrays. Next, some calculated examples are presented. These confirm that the input impedances of single dipoles can change significantly for strong coupling by surface waves. This shows up especially in the appearance of "blind angles" in radiation patterns of large arrays.

For strong coupling by surface waves it is obvious to ask whether elements of an array can be fed with the help of surface waves, as it is the case e.g. for a Yagi antenna. Simple considerations at the beginning of section 3.4.4 show that this is most possible for parallel dipoles, which are coupled by the TE_1- surface wave. To this end however, the dominant wave, i.e. the TM_1- surface wave, has to be suppressed. This is possible by an appropriate choice of the layer parameters. The remaining disposable degree of freedom is used to achieve a high excitation of the TE_1- surface wave as well as a radiation pattern as narrow as possible. In conclusion the lengths of five parallel dipoles are fixed with the help of very extensive numerical calculations so that the phase of all five antenna currents is nearly equal and hence broadside radiation is realized.

As a whole the investigations that have been carried out show that the technical applications of printed dipoles which have been effected so far have not nearly exhausted all possibilities. Especially promising seems to be the realization of special radiation patterns by a suitable choice of layer parameters, as well as the construction of radiation coupled arrays for application in the mm-wave range. Recently there has been an increased interest in the excitation of leaky waves, as with this type of wave it may be possible to design broadband antennas with a main lobe at an angle to the antenna surface.

Consideration of the structure with just two layers is extended to a structure with more layers and the location of the radiation elements in several layers, whereby more parameters are available for the design. The theory presented here is devised in a way that allows such extensions to be accommodated without the introduction of new principles.

CHAPTER 4
Focusing antennas for hyperthermia

4.1 Some remarks concerning the application of hyperthermia in medical therapy

In medicine hyperthermia of diseased tissue is being discussed as a possibility to treat tumors because these are particularly heat-sensitive, and because heat increases the effect of X-irradiation and anticancer drugs (for example [4.1] to [4.3]). A promising treatment with the help of hyperthermia requires antennas, which produce well controlled elevated temperature distributions in the tumor without overheating normal tissue.

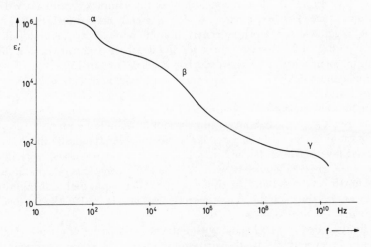

Fig 4.1 Real part of ε_r of muscle-like biological material [5.4]

The possibility that one can produce heat inside of dielectric material with the help of electromagnetic fields has been used for a long time in industry and medicine. If one - as is usually done - characterizes the electrical properties of the material which is to be heated with a complex dielectric constant, then the power converted to heat energy in a volume V is

$$P_V = \omega \int_V \varepsilon_0 \varepsilon_r'' |\vec{E}|^2 dV. \qquad (4.1)$$

Many investigations concerning the dielectric constants of biological tissues have been carried out by Schwan (for example [4.4]). Fig. 4.1 shows the approximate curve of ε_r' in the frequency range between zero and about 1 GHz. Three ranges which are based upon different relaxation mechanisms are characteristic: the reason for the first two, called α and β, is to be seen in the structure of the cells; the relaxation frequency of the γ-range is about the same as that of water. In connection with the problem posed here the range for approximately 100 MHz up to 10 GHz is of particular interest. For that purpose biological tissue may be classified into three major groups: material with very high water content (for example blood or cerebrospinal-fluid), tissue with moderate water content (for example skin or muscle) and material with low water content (for example, bones and fat). In Fig. 4.2 ε_r' and ε_r'' for tissue from all three groups are stated [4.5].

As a first result one can keep in mind that in the frequency range of approximately 1 GHz, $\varepsilon_r' \approx 50$ is valid for the first two groups and $\varepsilon_r' \approx 4$ for the third. In this range the power loss is about the same for all three groups: $\tan \delta \approx 0.3$. In detail one can conclude that the complex permittivity depends on temperature as well as on several physiologically caused factors.

If ε_r is known, the depth of penetration of the field can be determined analytically. Fig. 4.3 shows the curve for the depth of penetration as a function of the frequency based on the dielectric constants from Fig. 4.2. Note that electromagnetic fields of high frequencies can hardly penetrate into tissue of the groups 1 and 2.

The first calculations of the absorbed energy of electromagnetic waves in biological tissue were carried out within the simple model of a plane homogeneous wave normal to the tissue surface. It was assumed for the tissue that it consists of planar layers of skin, fat and muscle tissue (for example [4.6] to [4.12]). Because of the great difference between the dielectric constants of skin and muscle tissue on the one hand and fat tissue on the other hand, high reflections at the interfaces and thus standing wave patterns in the skin and in the fat layer were observed.

A limited area can be heated with antennas consisting of open-ended waveguides in direct contact with the surface of the tissue. In a paper of Guy [4.13] the field which is radiated from a rectangular waveguide into a bilayered planar tissue (fat- and muscle tissue) is determined. For the aperture field the transverse field of an undisturbed TE_{10}-mode is assumed. The radiation field of two open-ended waveguides, close to each other, is studied by Edenhofer [4.14]; the aperture field and the coupling for an excitation with a TE_{10}-mode is determined with the help

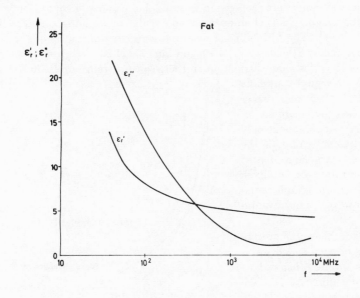

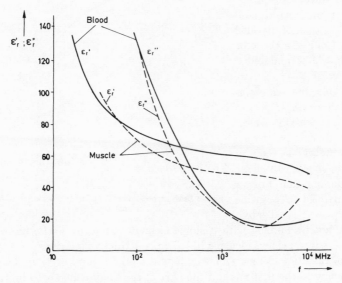

Fig. 4.2 Dielectric constant of biological material in the high frequency range

of the method of moments. Experimental results for open-ended rectangular- or circular waveguides are reported in several other papers ([4.15] to [4.17]). To reduce the fringing fields of an open-ended circular waveguide a corrugated flange is used [4.18]. The suitability of antennas consisting of multimode waveguides are analysed by Stuchly et al. [4.19] and Lin et al.[4.20]. While all investigations mentioned so far were carried out in the frequency range of about 1 GHz, the ridged waveguide antenna which is filled with water - applied by Paglione et al. [4.21] - operates at only 27 MHz. Because of the low frequency large depths of penetration are obtained; the large size of the radiator, however, is disadvantageous.

Besides the antennas consisting of open-ended waveguides several other radiating elements were studied: Coplanar waveguides [4.22], microstrip antennas ([4.23] to [4.26]), sets of parallel wires in front of a reflector [4.27], slots [4.28] and folded dipoles [4.29].

With all the antennas mentioned so far one can achieve penetration of electromagnetic energy into the tissue; however, the requirement which was stated at the beginning that only a well controlled area inside the body is heated obviously cannot be fulfilled with a single element applicator. The use of arrays suggests itself to achieve focusing based on constructive interference of the fields of several radiation elements. Two different methods have been used so far:

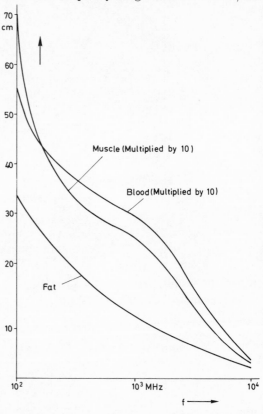

Fig. 4.3 Depth of penetration in biological material of uniform plane waves with normal incidence

According to the first method the part of the body which is to be treated is partly or totally surrounded by radiating elements ([4.30] to [4.37]). These are fed in such a way that the focus lies in the desired area. In the papers [4.30] to [4.33], results which were obtained for the frequency range of 50 to 150 MHz are given. Anderson et al. ([4.34] to [4.36]) used a homogeneous ellipsoid of lossy material as a model for the body and determined the field inside the body with the help of geometrical optics. Much more reliable theoretical results are obtained with the method outlined in [4.38], which is based on the solution of an integral equation ([4.33], [4.37], [4.39] and [4.40]). Here the non-homogeneous two-dimensional body is approximated by small rectangular subvolumes each of which is homogeneous. The field inside the body is determined by the method of moments. A realistic description of, for example, an arm makes a fine subdivision with respect to the wavelength necessary, and it thus leads to a very large system of equations. Therefore, that method is only applicable up to about 600 MHz. In [4.41] human limbs and necks are approximately represented by concentric cylindrical layers of fat, muscle and bone. The electric field in the tissue exposed to a plane wave is represented by an infinite series of cylindrical waves. The field of several radiators surrounding the body is approximated by superposing the solutions for several plane waves.

The second method is based on the before-mentioned stratified-layer model. This can be used advantageously in particular when heating of small areas in the muscle tissue not too deep under the surface is to be achieved. Gee et al. ([4.42] and [4.43]) determined the approximate current distribution of an array of 19 radiation elements, which are arranged parallel to the tissue surface, with the help of geometrical optics; the tissue is considered to be homogeneous. With the same assumption the field of 16 waveguide radiators is calculated in [4.44] starting with an aperture field, which is equal to the transverse field distribution of the TE_{10}-mode.

The equations used here (eqn. (2.17) together with Green's function which was determined in section 2.2) give an exact formulation of the relation between the current distribution and the field inside the layers (see also [4.45]). Another exact formulation is obtained using Weyl's approach [2.19] which describes the field of a current element as a superposition of plane waves. Thus, it is possible to reduce the problem to a two-dimensional Fourier transform (for example [4.46] and [4.47]). This method is advantageous when the field of an aperture not very small compared with the wavelength has to be determined. Furthermore, with this method one can state how to choose the field distribution of a plane aperture so that the electric field in a second plane parallel to this aperture has a focus with a maximum gain ("transverse focusing condition") ([4.48] and [4.49]). A general focusing condition applicable to arbitrarily orientated antennas has been derived in [4.50] and [4.51]. It tells how an array must be fed in order to get a maximum field strength in the focus for a given input power. This focusing condition is applied in [4.52].

All papers which have been quoted so far have in common that their radiating elements are located outside the body. In addition, methods have been studied with implantable antennas. Because the theoretical considerations are quite different from those concerning the model studied here, these investigations will not be addressed in the present book. A very detailed survey about all methods applied so far is given in a monograph of Hand and James [4.53].

The starting point for our considerations is the stratified-layer model. Dipole arrays will be used as radiating elements, the required current distribution of which can be achieved with the help of a suitable feeding network. In order to get a first insight, section 4.3.1 starts with considerations based on the reflection and transmission of plane homogeneous waves. Then some structures are investigated by using the method described in the chapter 2. It will be shown that the stratification of the medium complicates focusing. As preparation, we shall discuss in section 4.2 the results from published papers on focusing in homogeneous space.

4.2 Focusing antennas in free space

The use of the term focusing is particularly common in the field of geometrical optics. If it is applied, for example, to a typical radar antenna, then the antenna is said to be focused at infinity. In a paper which was published in 1949 [4.54], the field of a circular focusing aperture was analysed for the case of a focus within the Fresnel zone; that means that

$$z_f < 2\frac{d_A^2}{\lambda} \qquad (4.2)$$

(the aperture with the diameter d_A lies in the plane $z = 0$; z_f denotes the distance between the aperture plane and the plane with the focus; the focus is located on the axis of symmetry of the aperture). The phases of the electric currents were chosen for constructive interference, the amplitudes were constant. Examples of the axial fields show that this focus condition does not in every case lead to a maximum field strength in the focus. Eight years later Bickmore [4.55] determined the currents induced by an antenna of finite size in a second very small one, again under the condition that the second antenna lies within the Fresnel zone of the first antenna (inequality (4.2)). It can be seen that the induced currents have a maximum if the aperture of the transmitting antenna is part of a surface of a sphere and if the receiving antenna is located in the centre of this sphere with the focusing condition that constructive interference must be achieved. Sherman's paper [4.56] is also based on the condition of constructive interference; here examples are shown for rectangular plane apertures with various aperture fields. In a paper of Graham [4.57] the field of a focusing antenna is analysed as a function of the distance between the aperture and planes parallel to the aperture. Both the axial and the transverse field distributions are discussed by R.C. Hansen for tapered amplitudes of the aperture field [4.58].

In the five papers mentioned last it is assumed that the radiation field of each point source is proportional to e^{-jkr}/r, or in other words, that the observation points lie in the farfield of the point sources. The general focusing condition and also the "transverse" focusing condition are not subject to this restriction. For a plane aperture with circular symmetry in free space the latter requires that the aperture field be chosen proportional to

$$z_f \frac{e^{+jk\sqrt{x'^2 + y'^2 + z_f^2}}}{x'^2 + z'^2 + z_f^2} \left(-jk + \frac{1}{\sqrt{x'^2 + y'^2 + z_f^2}} \right) \qquad . \qquad (4.3)$$

If the observation point lies in the farfield of each point source, then the condition for the phase which has up to now been common is obtained. For observation points very close to the aperture the sum in eqn. (4.3) gives rise to an additional term. The new feature of eqn. (4.3) is that it also states a condition for the amplitude of the aperture field and not only for the phase.

4.3 Antennas above biological tissue

4.3.1 Reflection and transmission of plane uniform waves

The following ideas are based on a model of three layers (skin, fat and muscle tissue). The figures for the dielectric constants are taken partly from Fig. 4.2; the chosen frequency is 2.45 GHz (for further details concerning the choice of the frequency see 4.3.4.1). The thickness of the skin-layer is assumed as 3mm, that of the fat-layer as 5mm. As a characteristic feature of the stratified medium thus obtained (Fig. 4.4) one can point out the large difference between the dielectric constants of skin and muscle tissue on the one hand and that of fat tissue on the other hand.

As already mentioned several times before, the field of a Hertzian dipole can be represented as a superposition of plane waves according to Weyl's approach. Therefore, the propagation of plane waves will be discussed within the three-layer model. To simplify matters we shall assume plane uniform, transverse

ε_{r0}

Skin $\varepsilon_{r1} = 50 - j\,15$ $d_1 = 0{,}3\,cm$

Fat $\varepsilon_{r2} = 3{,}7 - j\,1{,}2$ $d_2 = 0{,}8\,cm$

Muscle $\varepsilon_{r3} = 50 - j\,20$

Fig. 4.4 Model of the tissue

electromagnetic waves. Neglecting the losses in the layers the ray tracings, shown in Fig. 4.5a for some angles of incidence ϑ_0, are obtained. The large differences of the dielectric constants of the layers lead to the following principal results:

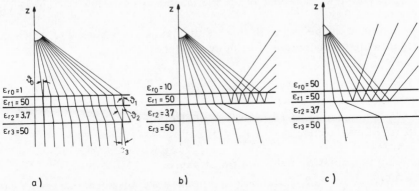

a)   b)   c)

Fig 4.5 Refraction and reflection of plane waves with oblique incidence upon lossfree 3-layered media

1) The angle ϑ_3 is only slightly bigger than zero; this means that the rays inside the muscle run almost vertically to the interfaces, independent of the angle of incidence.

2) For

$$\vartheta_i = \operatorname{arctg}\sqrt{\frac{\varepsilon_{r(i+1)}}{\varepsilon_{ri}}} \tag{4.4}$$

the reflection coefficient of the TM-wave at the corresponding interface is zero (Brewster angle).

In order to get a focus, different rays must intersect in one point according to geometrical optics. Because of the fact that, according to 1), all rays run almost parallel to each other inside the muscle, the possibility of producing a focus inside the muscle is limited. However, different values for ϑ_3 are obtained by embedding the sources in a material with a high dielectric constant (Figs. 4.5b and 4.5c). The disadvantage of this procedure is, however, that then total reflection occurs, thus causing the loss of part of the power radiated into the medium.

To obtain more detailed knowledge including the effect of the Brewster angle, of the total reflection and of the losses in the three layers, in Fig. 4.6 the power $S_z^{TM/TE}$ transmitted into layer three by an incident wave with $S_{zin}^{TM/TE}$ as a function of the angle of incidence and for various rates of ε_{r0} is shown. One can see that the angular spread available increases as ε_{r0} decreases, while at the same time the transmitted power becomes smaller in the case of small angles of incidence. This decrease is obviously caused by high reflections because of $|\varepsilon_{r0}| \ll |\varepsilon_{r1}|$. A better adjustment of the source to the tissue can be obtained by dividing the "0"-layer up into several layers, that means by using matching layers as known from the transmission line theory. Fig. 4.7 shows some examples with the parameters specified in such a way that for vertical incidence the reflection coefficient is zero. One can see that by means of a suitable choice of ε_{r0} the angular spread which is used for the transmission of power into the muscle layer can be specified.

Almost as simple is the method of estimating the influence of a reflector mounted above the sources. Obviously a distance of

$$d_s = \frac{n_s \lambda}{4\sqrt{\varepsilon_{r0}}}, \qquad n_s = 1, 3, 5 \cdots \tag{4.5}$$

leads to an addition of the contribution of the source itself and of its image source in the direction of $-z$.

4.3.2 The radiation field of a point source

Based on section 2.4.1 the analytical properties of the integrands essential for the solution of the Sommerfeld integrals can be stated. When we have a device with a reflector the branch points lie at

$$k_{\rho b1} = \pm k_3 = \frac{2\pi}{\lambda}\sqrt{\varepsilon_{r3}} \,, \tag{4.6}$$

when we have an unscreened device there are the following branch points in addition:

$$k_{\rho b2} = \pm k\sqrt{\varepsilon_{r0}} \qquad . \tag{4.7}$$

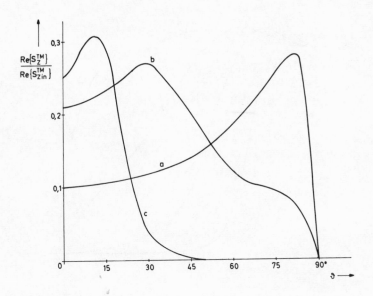

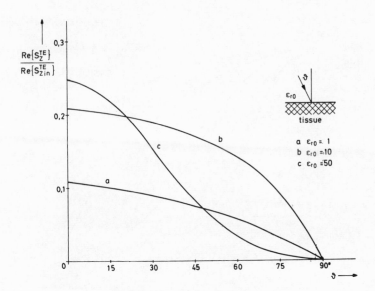

Fig. 4.6 Penetration of plane uniform waves into the muscle layer
(S_z = z - component of the Poynting vector)

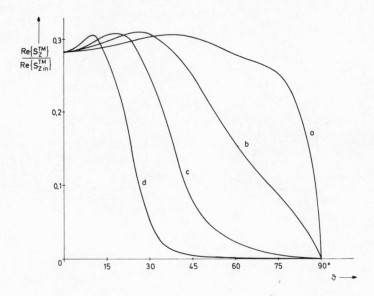

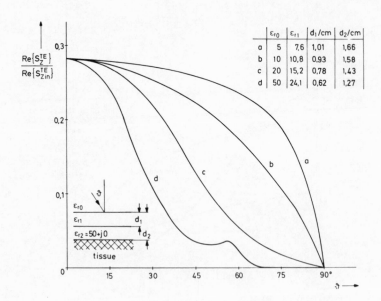

Fig. 4.7 Penetration of plane uniform waves into the muscle layer
improved by transmission layers

Undamped surface waves cannot occur because of the losses in the tissue, therefore the reflection coefficients $\Gamma_0^{TM/TE}$ have no poles on the real k_ρ-axis.

In the case of a screened device it is not possible to solve the Sommerfeld integrals with the help of the saddle-point method because of the strong oscillations of the integrands, caused by the term $S_\pm^{TM/TE}$ (eqns. (A3.27) and (A3.28)), occurring near the saddle point; this means that the requirements stated in section 2.4.2 are not fulfilled. Also in the case of an unscreened device the use of the saddle-point method is not sensible in most cases, because first of all, in a rather complicated procedure, poles in the complex k_ρ-plane have to be determined. All results shown in the following sections were therefore obtained by numerical integrations. Only when the method of moments is applied the approximate solutions (2.174) to (2.178) for very small distances between source and observation point are used in addition as described in Appendix A1.

For the representation of the results, lines of constant level are used for the amplitudes of the field components and lines of constant phase for the phases (for the construction of the field pattern a differential equation would have to be solved in addition).

The components of the electric field are illustrated in the range $z < -d_0$; dE_ρ and dE_z in the xz-plane, dE_φ in the yz-plane. By way of multiplying with $\cos\varphi$ or $\sin\varphi$ respectively, (compare e.g. eqns. (A3.47) to (A3.58)) and superposition, the total field is obtained.

As a first example the field for the same structure as in Fig. 4.6c with a reflector at $d_s = \lambda_0/4$ and with $d_0 = 2$ cm is shown in Fig. 4.8. The following details are important:

1) $|dE_z/\cos\varphi\,|$ in muscle tissue is much smaller than $|dE_\rho/\cos\varphi\,|$ and $|dE_\varphi/\sin\varphi\,|$. This fact is in line with the almost vertical run of the rays, as shown in Fig. 4.5c (the range $\rho \leq 0.25$ cm is excluded in the representation of $|dE_z/\cos\varphi\,|$ because of $|dE_z(\rho = 0)| = 0$). However, $|E_z/\cos\varphi\,|$ in the fat layer at $\rho \approx 2$ cm is approximately equal to the maxima of $|E_\rho/\cos\varphi\,|$ and $|E_\varphi/\sin\varphi\,|$, which are located in the skin layer just beneath the source. This corresponds with the almost horizontal run of the rays in the fat layer shown in Fig. 4.5c.

2) Inside the tissue $|dE_\rho/\cos\varphi\,| \approx |dE_\varphi/\sin\varphi\,|$ for $\rho \leq 4$ cm.

3) Because of the losses inside the tissue and because of the reflections at the interfaces $|dE_\rho|$ and $|dE_\varphi|$ decay strongly as the distance from the source grows. Thus, e.g. for the two field points $P(z = -d_0, \rho = 0)$ (skin surface) and $P(z = -d_2, \rho = 0)$ (interface fat-muscle tissue) the levels differ as

$$D_{02\rho} = 20\lg\left|\frac{dE_\rho(z = -d_2, \rho = 0, \varphi = 0°)}{dE_\rho(z = -d_0, \rho = 0, \varphi = 0°)}\right| = -9.4 \text{ dB} \qquad , \qquad (4.8)$$

$$D_{02\varphi} = 20\lg\left|\frac{dE_\varphi(z = -d_2, \rho = 0, \varphi = 90°)}{dE_\varphi(z = -d_0, \rho = 0, \varphi = 90°)}\right| = -9.4 \text{ dB} \qquad . \qquad (4.9)$$

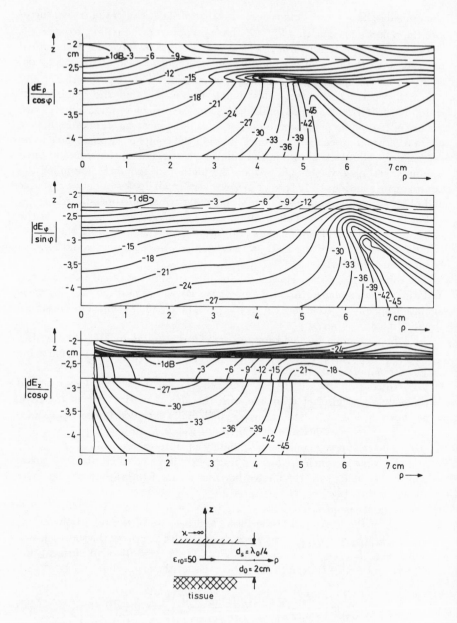

Fig. 4.8a Electrical field of a point source inside the tissue (amplitudes)

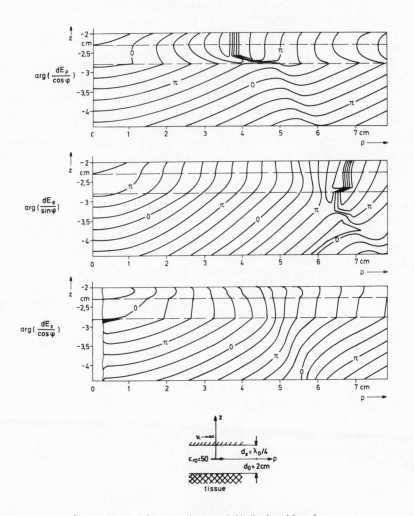

Fig. 4.8b Electrical field of a point source inside the tissue (phases)

Thus, the values obtained for dE_ρ and dE_φ are the same.

4) At approximately $\rho = 6$ cm, $|dE_\rho|$ and $|dE_\varphi|$ have a distinct minimum.

5) The curves of constant phase for dE_ρ and dE_φ run in the muscle tissue for $\rho \approx$ 0 parallel to the interfaces, for $\rho \geq 2$ cm they run at an angle of approximately $25° - 40°$. Thus, obviously only the range $\rho \leq 2$ cm can roughly be described by means of the considerations of section 4.3.1.

Altogether it is apparent from Fig. 4.8 that results based on tissue models consisting of homogeneous media can hardly be transferred to stratified media.

Fig. 4.9 shows the dE_ρ- and dE_φ-component of a point source embedded in a material with a low dielectric constant ($\varepsilon_{r0} = 10$). We notice the smooth curves for the amplitudes and the phases. Inside muscle tissue the lines of constant phase - in agreement with the explanations of section 4.3.1 - run less steeply than in Fig. 4.8. For the level differences defined in eqns. (4.8) and (4.9) the more favourable values

$$D_{02\rho} = D_{02\varphi} = -7.6 \text{ dB} \tag{4.10}$$

are now obtained. The widths of the main lobes are described by stating the ρ-values for which the amplitude of a field component in the fat-muscle interface is 3 dB smaller than for $\rho = 0$. This value will be called b_g, with a further index denoting the field component which the stated value is based on. For the examples mentioned so far the following values are obtained:

Fig. 4.8

$$b_{g\rho} = 1.9 \text{ cm} \qquad , \tag{4.11}$$

$$b_{g\varphi} = 1.6 \text{ cm} \qquad , \tag{4.12}$$

Fig. 4.9

$$b_{g\rho} = 2.4 \text{ cm} \qquad , \tag{4.13}$$

$$b_{g\varphi} = 2.0 \text{ cm} \qquad . \tag{4.14}$$

In Fig. 4.10a and 4.10b only the amplitudes of the E_ρ-components for the structure of Fig. 4.7 cases a) and d), are plotted. As was to be expected the penetration depth into the structure according to Fig. 4.10a is bigger than that according to Fig. 4.10b. The values for the width of the main lobe are almost the same. The results can be specified as follows:

Fig. 4.10a

$$b_{g\rho} = 1.8 \text{ cm} \qquad , \tag{4.15}$$

$$D_{02\rho} = -7.3 \text{ dB} \qquad , \tag{4.16}$$

Fig. 4.10b

$$b_{g\rho} = 1.9 \text{ cm} \qquad , \tag{4.17}$$

$$D_{02\rho} = -8.8 \text{ dB} \qquad . \tag{4.18}$$

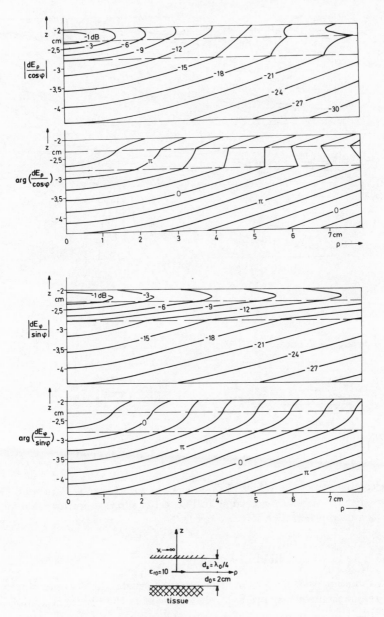

Fig. 4.9 Electrical field of a point source inside the tissue

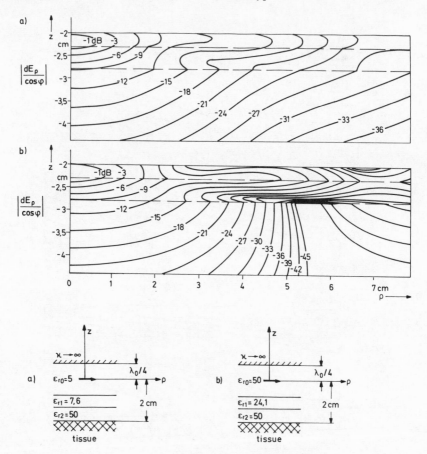

Fig. 4.10   Electrical field of a point source inside the tissue

A comparison of the values (4.8) and (4.11) with those of (4.15) and (4.16) shows that the structure according to Fig. 4.10a allows a narrow main lobe as well as a deep penetration into the tissue.

4.3.3 The radiation field of a dipole

As radiation elements for the focusing antennas, flat dipoles operating at the first resonance frequency are used. It follows from the results of section 3.3.1 that then one has to put

$$2l \approx \frac{\lambda}{2\sqrt{\varepsilon_{r0}}} \tag{4.19}$$

Then the current distribution is approximately sinusoidal (eqn. (3.97)) and

$$\lambda_e \approx \frac{\lambda}{\sqrt{\varepsilon_{r0}}} \quad , \tag{4.20}$$

further the real part of the input impedance is approximately

$$Re\{Z_{in}\} \approx \frac{73\Omega}{\sqrt{\varepsilon_{r0}}} \quad . \tag{4.21}$$

The exact results for the resonance length, the current distribution and the input impedance are determined numerically. Based on the current distribution thus obtained in the following, the field of a $\lambda_e/2$ flat dipole is calculated from eqn. (2.17). If the field components E_ρ and E_φ which have so far been used are converted into E_x and E_y, one obtains for the main lobe

$$|E_x| \gg |E_y| \quad . \tag{4.22}$$

Further, as already valid for the current element, it is valid also that

$$|E_x| \gg |E_z| \quad . \tag{4.23}$$

The E_x-component dominates also outside the main lobe in the skin layer and in the muscle tissue, but $|E_x|_{max} \approx |E_z|_{max}$ in the fat layer. As an example Fig. 4.11 shows the amplitudes of all three components in the plane $z = -4$ cm, which means 12 mm below the fat-muscle interface for the structure of Fig. 4.8 ($\varepsilon_{r0} = 50$, $d_s = 0.433$ cm). The following values can be read from Fig. 4.11:

$$
\begin{aligned}
\text{maximum side lobe level of } E_y : &\quad \approx -9 \text{ dB} \\
\text{maximum side lobe level of } E_z : &\quad \approx -15 \text{ dB} \quad .
\end{aligned}
\tag{4.24}
$$

4.3.4 The radiation field of an array of dipoles

4.3.4.1 Some basic considerations

First of all we shall fix a criterion for the quality of a focusing antenna. This criterion has to take into account the principal problem that occurs during the application of hyperthermia: This consists in the danger that the skin layer is scorched if a certain area inside a muscle is heated sufficiently. We shall prefer antennas with a large ratio of $|\vec{E}|$ in the focus to the maximum of $|\vec{E}|$ in the skin layer. This ratio shall be denoted by D_f. Formally D_f is determined by the equation

$$D_f = 20 \lg \frac{|\vec{E}| \text{ in the focus}}{|\vec{E}|_{max} \text{ in the skin}} \quad . \tag{4.25}$$

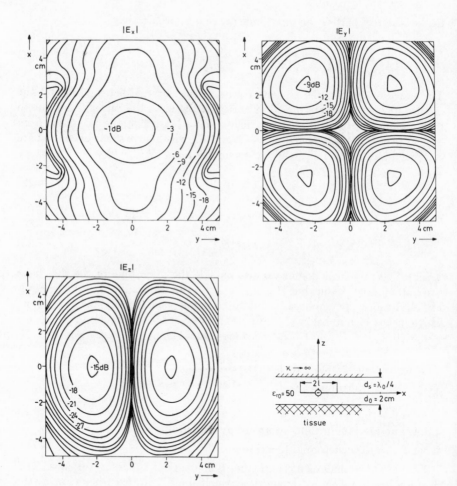

Fig. 4.11a Electrical field of a $\lambda_e/2$-dipole in the plane $z = -4$cm (amplitudes)

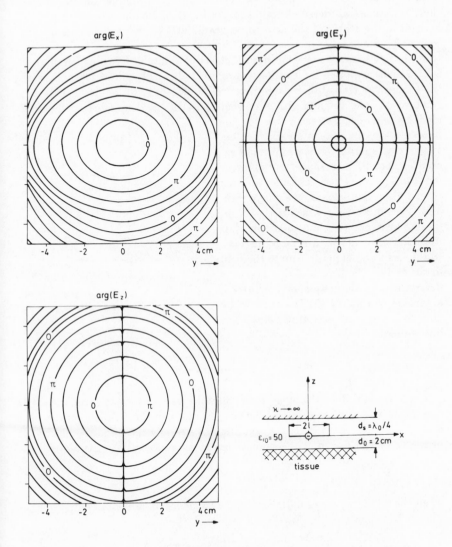

Fig. 4.11b Electrical field of a $\lambda_e/2$ - dipole in the plane $z = -4\,cm$ (phases)

Similarly, the 3–dB beam width of the main lobe in the plane of the focus is denoted by b_f. The second index is x or y, according to whether the xz–plane or the yz–plane is considered. Also in those cases, in which a non-focusing antenna is applied, the centre of the area to be heated will be called the focus in order to simplify matters.

In the following we shall consider the choice of a suitable frequency in more detail. Our discussion so far leads us to two contradicting requirements concerning the frequency: One obtains large penetration depths with low frequencies and small sizes of the antenna with high frequencies. If one demands that the stratified layer model be reasonable for the human body, the maximum distance between two radiation elements must not exceed 7 – 9 cm. The arrangement of larger plane arrays is not very useful for another reason also: the fields of the single radiation elements strongly decrease with increased distance to the source and thus radiation elements with a large distance to the focus do not contribute to the field in the focus. The length of the dipoles is determined by the frequency and by ε_{r0} according to eqn. (4.19). In order to be able to arrange at least three dipoles in line we require that

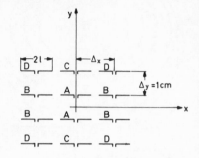

Fig. 4.12 Array of 12 dipoles

$$0.5\lambda_0 = 0.5\frac{\lambda}{\sqrt{\varepsilon_{r0}}} < 3 \text{ cm} \quad . \tag{4.26}$$

In the Federal Republic of Germany only the frequencies 434 MHz, 2.45 GHz and 4.8 GHz are allowed for medical applications in the frequency range 100 MHz to 10 GHz. Because of the condition (4.26) only the frequency 2.45 GHz proves to be suitable, if at the same time the values for ε_{r0} are limited by

$$\varepsilon_{r0} \gtrsim 5 \quad . \tag{4.27}$$

Then the following wavelengths inside the tissue layers are obtained:

$$\begin{aligned}
\lambda_1 &= 2\pi/Re\{k_1\} = 1.71 \text{ cm} &\quad \text{skin,} \\
\lambda_2 &= 2\pi/Re\{k_2\} = 6.28 \text{ cm} &\quad \text{fat,} \\
\lambda_3 &= 2\pi/Re\{k_3\} = 1.64 \text{ cm} &\quad \text{muscle.}
\end{aligned} \tag{4.28}$$

For the construction of the array, 12 flat dipoles, fed by a suitable network are arranged in a rectangular grid (Fig. 4.12). The spacing between two lines of dipoles, Δy, is in all cases discussed here 1.0 cm, the spacing between two columns, Δx, is 3.0 cm for $\varepsilon_{r0} = 5.0$, 2.5 cm for $\varepsilon_{r0} = 10.0$ and 2.0 cm for $\varepsilon_{r0} = 5.0$. The centre of the area to be heated – the focus – is located at $\rho = 0, z = -4.0$ cm, which means 2.0 cm beneath the surface of the skin. The currents on all dipoles and the input impedances are calculated by the solution of the complete boundary value problem, thus taking into account the mutual coupling of the dipoles, too.

4.3.4.2 Uniform current distribution

In a first step the electrical field was calculated for a uniform current distribution. The results are discussed with the aid of the structures used in the preceding sections. In order to facilitate orientation we provide the following list of the figures. The different structures are numbered using Roman numerals, and those figures which refer to the same structure are listed in the corresponding line.

structure	
I	4.6c; 4.8a+b; 4.11a+b; 4.13a-c; 4.16a
II	4.6b; 4.9; 4.14b; 4.17
III	4.7d; 4.10b; 4.14a; 4.16c
IV	4.7a; 4.10a; 4.14a; 4.15; 4.16d

Figs. 4.13a and b show $|E_x|$ and $|\vec{E}|$ in the xz–plane. We have a maximal field strength in the skin layer for $\phi = 0°$, $\rho \approx 1.3$ cm and a very low field strength (about -8 dB below the maximum) in the area directly beneath the centre of the array. As Figs. 4.8a and b for the Green's function do not show such a minimum, this must be achieved by interference. The contours for $|E_x|$ and $|\vec{E}|$ in the xz–plane are very similar, the same is true for the yz–plane. Obviously $|E_z|$ and $|E_y|$ are very low in all cases.

In order to calculate the power absorbed per volume, denoted $|\tilde{E}|$, a multiplication with ε_{ri}'' must be carried out according to eqn. (4.1). In the logarithmic representation chosen here, only the value for $10\log(\varepsilon_{ri}'')$ must be added to the levels given before. For the three layers which are of interest here these values are

$$
\begin{array}{lll}
\text{skin} & 10\log(\varepsilon_{ri}'') = 11.8 \text{ dB} & \\
\text{fat} & \qquad'' \qquad = 0.80 \text{ dB} & (4.29) \\
\text{muscle} & \qquad'' \qquad = 13.0 \text{ dB} &
\end{array}
$$

Fig. 4.13c shows $|\tilde{E}|$ which belongs to the field distribution of Fig. 4.13b. As was to be expected, both figures look very similar in the skin and in the muscle layers, the level in the fat layer however drops down because of the low value of ε_r''. If by analogy with eqn. (4.25)

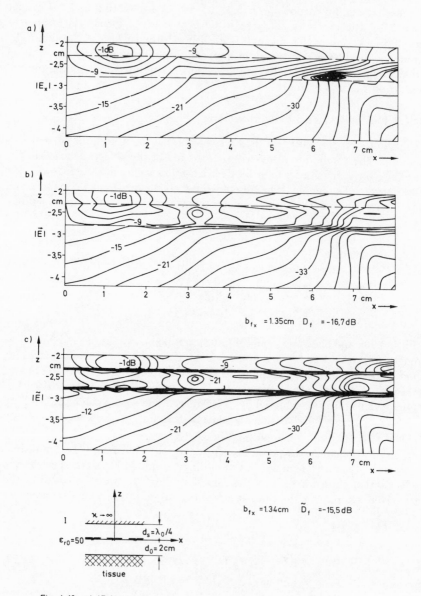

Fig. 4.13 a) $|E_x|$
        b) $|\vec{E}|$ of an array (see fig. 4.12) with uniform current distribution
        c) $|\tilde{E}|$

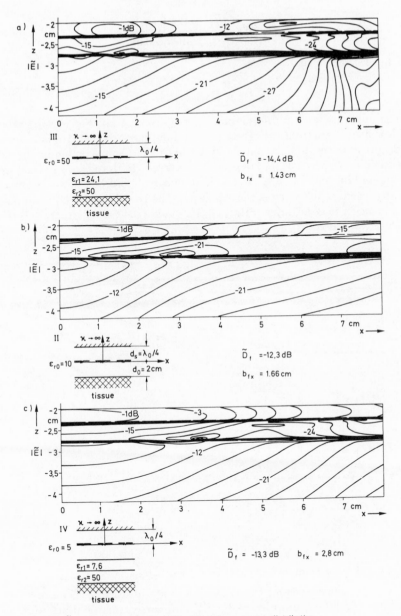

Fig. 4.14 $|\tilde{E}|$ of an array (see fig. 4.12) with uniform current distribution

$$\tilde{D}_f = 20 \lg \frac{|\tilde{E}| \text{ in the focus}}{|\tilde{E}|_{max} \text{ in the skin}} \quad , \tag{4.30}$$

is defined, here we get

$$\tilde{D}_f = D_f + 1.2 \text{ dB}. \tag{4.31}$$

The application of transformation layers (structure III, Fig. 4.14a) avoids the low field strength in the centre area and improves the penetration of electromagnetic waves into the muscle by about 1 dB. The maximum in the skin layer at $\phi = 0°$, $\rho = 1.3$ cm, however, is preserved. Both remaining structures, II (Fig. 4.14b) and IV (Fig. 4.14c), have maximal absorbed power below the centre element of the array in the skin layer. The best result for $\tilde{D}_f$ is obtained for a homogeneous "0"-layer with $\varepsilon_{r0} = 10.0$ above the tissue (structure II, Fig. 4.14b, $\tilde{D}_f = -12.3$ dB).

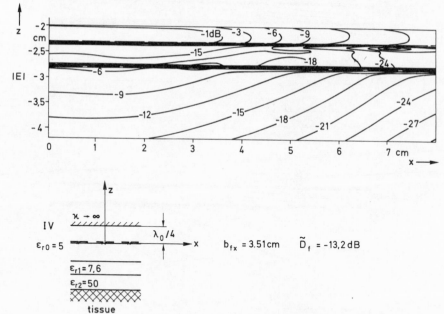

Fig. 4.15 $|\tilde{E}|$ of an array (see fig. 4.12) each dipole fed with 1V

The input impedances of the 12 dipoles is not identical because of mutual coupling. Therefore, in order to get uniform current distribution, the input voltages have to be adjusted. The following table gives as an example the input voltages which belong to the plots of Figs. 4.13a-c (The number of the dipoles is given in Fig. 4.12).

dipoles	input voltage
A	$(7.3 + j1.8)$V
B	$(7.5 + j1.5)$V
C	$(10.6 + j5.9)$V
D	$(10.2 + j5.2)$V

If this adjustment is not done, in order to avoid additional effort, slightly different results are obtained as can be seen e.g. by comparison of Fig. 4.15 and Fig. 4.14c (both structure IV).

4.3.4.3 Application of focusing arrays

The general focusing condition as formulated in [4.51] is based on the conjugate–field matching concept. For its explanation, it is assumed that a point source is located in the focus. At this time the array of dipoles is used as a receiving antenna. The received voltages are (V). In a next step the array is fed. The general focusing condition guarantees a maximum field strength in the focus for a given input power if the input currents are chosen as

$$(I) = (Z)^{-1} \cdot (V)^* \qquad (4.32)$$

where (Z) is the impedance matrix of the array. This equation is exact including the mutual interaction between the dipoles. Both, (Z) and (V) can be determined by solving the integral eqn. (2.21) and applying the reciprocity theorem. The physical interpretation of eqn. (4.28) is, that the phases of the currents must be adjusted for constructive interference and the amplitudes must decrease with increasing distance to the centre of the array. Many calculations show that the specification concerning the amplitudes of the currents is not critical here, therefore the plots we present have been made with constant amplitudes. The phases of the currents for constructive interference can be obtained directly from the plots of the lines of constant phases for one dipole only (see e.g. Fig. 4.11b). As $|E_x| \approx |\vec{E}|$ in the muscle tissue, the phases of E_x are used. The voltages which are required for these currents are determined by solving the integral eqn. (2.21) thus taking into account the effect of the mutual coupling and of the stratification.

Figs. 16a-d give the absorbed power and the values of $\tilde{D}_f$ as well as of b_{fx} which are obtained if the general focusing condition is applied. Comparing these results with those for uniform current distribution (Figs. 4.13c and 4.14a-c) shows, that the general focusing condition, although obvious at first sight, does not automatically lead to better results for hyperthermia. For the structure II (Fig. 4.14b / Fig. 4.16b) we now get an even smaller value for $\tilde{D}_f$. The reason is that, as already mentioned, the general focusing condition only guarantees maximal field strength with respect to the total radiated power P_s (the antennas

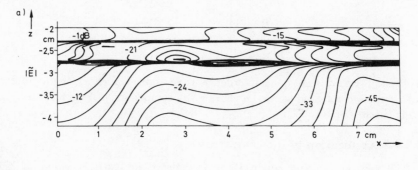

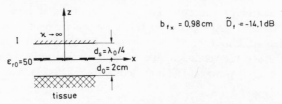

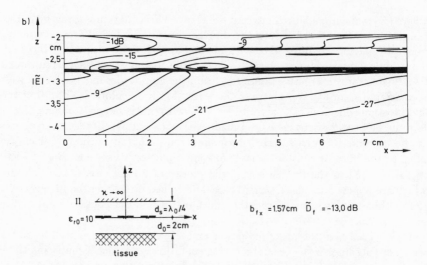

Fig. 4.16 a + b $|\tilde{E}|$ of an array (see fig. 4.12) with feeding according to the general focusing condition

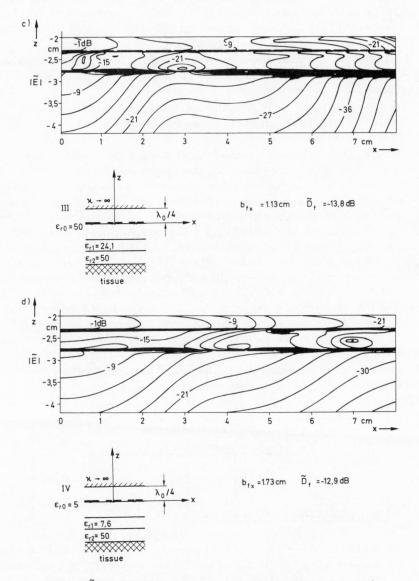

Fig.4.16 c+d I$\tilde{E}$I of an array (see fig. 4.12) with feeding according to the general
focusing condition

are supposed to be lossless). If the phases of the currents are varied, the input impedances and, by this, P_s are varied, too. Thus we may get a worse $\tilde{D}_f$ in spite of an optimal gain. Further as it is not a critical point to generate the microwave power that is required for hyperthermia, the application of the general focusing condition is not often much help.

Instead of this we wish to specify an array with a $\tilde{D}_f$ as large as possible. It should be remembered that many more parameters are available within the concept outlined here than have been discussed so far:

- The spacing between the dipoles; this spacing is not necessarily uniform
- the spacing between the array and the surface of the skin
- the spacing between the array and the reflector.

As the analysis of one setup even with a high speed computer (e.g. a vector computer) is extremely time-consuming, it is completly impossible to carry out an optimization procedure which takes into account all available parameters. Therefore the following discussion is restricted to only one structure, structure II, the analysis of which is less expensive because of the homogeneous layer $+\hat{d}_0 \leq z \leq -d_0$ ($\varepsilon_{r0} = 10.0$) and which led to good results (see Fig. 4.14 b).

It is not possible to determine the optimal phases of the antenna currents by analytical or by simple approximate methods but only by systematic variation. It proved to be sufficient to vary the phase in steps of 30°. The phases of the two centre dipoles A (see Fig. 4.12) were set to 0°. Then because of the symmetrical setup of the array only the phases of dipoles B, C and D had to be varied in steps of 30°. This results in 1728 different cases which had to be investigated. In each case $|\tilde{E}|$ was calculated in the skin layer in the range $0 \leq \rho \leq 3$ cm and in the focus ($\rho = 0$, $z = -4$ cm). We got an optimal value of $\tilde{D}_f = -11.0$ dB for the phases given in the table below. Additionally the phases and $\tilde{D}_f$ according to the general focusing condition are stated.

feeding	phases of dipoles				
of the dipoles	A	B	C	D	$\tilde{D}_f$
according to the general focusing condition	0°	60°	30°	90°	-13.0
for optimal $\tilde{D}_f$	0°	−30°	−120°	−90°	-11.0

We see that both procedures lead to completely different phases. The value for $\tilde{D}_f$ is improved by 2.0 dB. Fig. 4.17 gives the electrical field distribution, the corresponding plot for the general focusing condition is given in Fig. 4.16b. Inside the muscle tissue both plots are very similar except for a shift of the contours by the difference of the values for $\tilde{D}_f$. Inside the skin layer however the 3–dB beam width is about 80 mm in the case of optimal $\tilde{D}_f$ compared with about 40 mm in the case of the general focusing condition. This means that the development of a critical small hot spot is avoided.

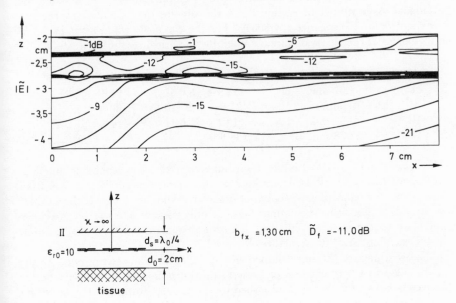

Fig. 4.17 $|\tilde{E}|$ of an array (see Fig. 4.12) with feeding for optimal $\tilde{D}_f$

In order to find out how successful the procedures used so far have been in increasing the penetration depth of electromagnetic fields into the muscle layer, we also take values for $\tilde{D}_f$ from the field patterns of the current elements. As a reference we choose the power absorbed per volume in the observation point which is situated 12 mm below the interface between muscle and fat tissue ($\rho = 0$). The following list displays the values which we have obtained. In addition the results from Fig. 4.12 are listed for the sake of comparison.

structure	point source	general focusing condition	phases due to achieve optimum $\tilde{D}_f$
I	$\tilde{D}_f = -19.3$ dB	$\tilde{D}_f = -14.1$ dB	
II	$\tilde{D}_f = -17.0$ dB	$\tilde{D}_f = -13.0$ dB	$\tilde{D}_f = -11.0$ dB
III	$\tilde{D}_f = -18.0$ dB	$\tilde{D}_f = -13.8$ dB	
IV	$\tilde{D}_f = -16.3$ dB	$\tilde{D}_f = -12.9$ dB	

They show that at best (structure II with phases for optimized $\tilde{D}_f$ to structure I with a point source) an improvement of 8.3 dB could be achieved.

4.3.4.4 Matching of the antennas

For practical use considerable mismatch of the antenna elements should be avoided. As outlined in section 4.3.3 the dipoles are operated at the approximately first resonance frequency. For the cases discussed here the input impedances of the single dipoles obtained by numerical resolution are

structure	dipole length	ε_{r0}	$(85.3 + j39.5)\Omega/\sqrt{\varepsilon_{r0}}$	input impedance
I	8.67 mm	50	$(12.1 + j5.6)\Omega$	$(14.6 + j6.2)\Omega$
II	19.36 mm	10	$(27.0 + j12.5)\Omega$	$(30 + j13.3)\Omega$
III	8.67 mm	50	$(12.1 + j5.6)\Omega$	$(14.8 + j6.9)\Omega$
IV	27.38 mm	5	$(38.6 + j17.7)\Omega$	$(42.4 + j21.6)\Omega$

The value $(85.3 + j39,5)\Omega$ is the input impedance of a corresponding dipole in free space ($\varepsilon_r = 1$). It is determined by numerical procedure or taken from any antenna engineering textbook. Comparing columm 4 and 5 shows that, as already stated, the input impedance is only weakly affected by interfaces not close to the source. This fact is also stressed by the very similar input impedances obtained for structures I and III.

The real part of the input impedance specifies the radiated power. As the setup suggested here consists of lossless transformation layers and a reflector above the array, this power is absorbed by the tissue. However, high values for ε_{r0}, which are often suggested for the development of compact applicators, lead to low input impedances and thus the efficiency deteriorates. The real part is even lower, if the antennas are chosen shorter than a quarter of a wavelength in order to obtain a real input impedance.

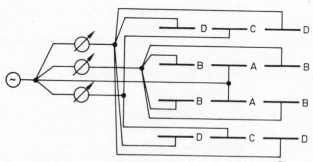

Fig. 4.18 Feed circuit of the array of 12 dipoles

The setup of 12 dipoles favours a feed circuit as shown in Fig. 4.18, consisting of 5 power dividers, three phase shifters and if necessary three attenuators. This leads, in any case, to even lower values for $Z_{in}^{(1)}$, the input impedance of the four power lines. Altogether, with respect to the input impedance a low ε_{r0} (e.g. $\varepsilon_{r0} = 10.0$, as chosen for structure II) is more advantageous than the often preferred value $\varepsilon_{r0} = 50.0$.

The input impedances of the dipoles are strongly affected by the mutual coupling. The table gives as an example the mutual coupling between the dipole

dipole	structure			
	I	II	III	IV
1,1	$(15.1 + j6.6)\Omega$	$(28.5 + j12.6)\Omega$	$(15.2 + j6.8)\Omega$	$(42. + j17.3)\Omega$
1,2	$(-1. + j.5)\Omega$	$(2.7 - j6.5)\Omega$	$(-1.2 + j.1)\Omega$	$(16.5 - j.7)\Omega$
1,3	$(-.7 - j.4)\Omega$	$(-1. + j2.8)\Omega$	$(.3 - j.7)\Omega$	$(-3.7 - j1.1)\Omega$
2,1	$(-3.4 - j5.5)\Omega$	$(18.9 - j14.7)\Omega$	$(-2. - j3.1)\Omega$	$(38.6 - j1.1)\Omega$
2,2	$(0. + j1.1)\Omega$	$(-1.8 - j7.8)\Omega$	$(.6 + j.6)\Omega$	$(11.5 - j8.6)\Omega$
2,3	$(-.6 - j.2)\Omega$	$(-.1 + j2.7)\Omega$	$(.2 - j.5)\Omega$	$(-4.4 - j.2)\Omega$
3,1	$(.2 + j2.8)\Omega$	$(-5.9 - j16.8)\Omega$	$(3.5 + j.6)\Omega$	$(16.2 - j19.5)\Omega$
3,2	$(.3 - j.1)\Omega$	$(-8.2 - j1.6)\Omega$	$(-.5 - j1.5)\Omega$	$(.4 - j12.5)\Omega$
3,3	$(-.4 - j.1)\Omega$	$(2.3 + j2.3)\Omega$	$(.3 - j.6)\Omega$	$(-4.6 + j1.7)\Omega$
4,1	$(-.3 + j.5)\Omega$	$(-15.9 + j.5)\Omega$	$(-4. - j2.1)\Omega$	$(-3.2 - j17.9)\Omega$
4,2	$(.4 + j.4)\Omega$	$(-6. + j8.)\Omega$	$(-1. + j1.9)\Omega$	$(-7.9 - j8.)\Omega$
4,3	$(-.5 + j.2)\Omega$	$(4.3 + j.4)\Omega$	$(-.7 - j.8)\Omega$	$(-3.4 + j3.4)\Omega$

located in the upper left corner (dipole 1,1) and the other 11 dipoles, and in addition, in the first line, the self-impedances, for the structures I – IV. (The numbers of a dipole give the number of its row and of its column, starting with the left upper corner in Fig. 4.12.) As is to be expected, the coupling of parallel dipoles is much stronger than of dipoles in line. The magnitude of the maximal mutual coupling is up to about 85% of the magnitude of the self-impedances. This fact makes the matching much harder. Changing the mode of operation results in very different input impedances. As an example the next table gives the input impedances for the current distributions discussed so far (only structure II). From the table, it follows that changing the mode of operation requires a change of the matching circuits.

dipoles	feeding of the array			
	uniform voltage	uniform current	according to the general focusing condition	for optimal $\tilde{D}_f$
A	$(23.9 - j25.6)\Omega$	$(28.3 - j23.8)\Omega$	$(71.3 - j28.8)\Omega$	$(9.8 + j5.7)\Omega$
B	$(30.2 - j22.7)\Omega$	$(36.5 - j20.2)\Omega$	$(44.5 - j19.8)\Omega$	$(25.0 - j12.2)\Omega$
C	$(42.1 - j8.9)\Omega$	$(23.3 - j1.6)\Omega$	$(8.2 - j1.0)\Omega$	$(50.4 + j32.2)\Omega$
D	$(42.3 - j1.0)\Omega$	$(25.7 - j1.3)\Omega$	$(15.4 + j0.0)\Omega$	$(50.3 + j7.1)\Omega$

4.4 Summary of Chapter 4 and some further comments

Hyperthermia with electromagnetic fields has proved to be a promising tool in tumor treatment, expecially if it is combined with X–irradiation and anticancer drugs. The theoretical determination of the field inside the body must be based on a careful measurement of the dielectric constants of biological tissue. Roughly this can be classified into three groups: Material with high, moderate and low water content. High water content results in high values for ε_r'' and, with this, in low penetration depth for microwave frequencies.

As applicators, above all various types of waveguide antenna have been investigated, and to some extent also microstrip antennas, dipoles, slots, reflector antennas etc. If deeply located areas are to be treated, the body is partly or totally surrounded by the antennas. In this book the heating of small areas not deep beneath the surface is discussed; therefore, the stratified layer model can be applied.

The aim is to heat a limited and well defined area inside the body. It is obvious that focusing antennas should be applied. Therefore, as a preliminary, in section 4.2 focusing antennas in free space are briefly discussed. In section 4.3 the field distributions produced by dipoles inside the tissue are investigated in detail. As an introduction, in section 4.3.1 the reflection and transmission of uniform plane waves with oblique incidence upon layered media are discussed, showing e.g. the effect of the Brewster angle and the total reflection. High reflections at the surface are reduced if matching layers are used. Comprehensive information about the penetration of electromagnetic waves into the tissue is obtained by calculating the field of a point source (section 4.3.2); that means by determining the Green's function of the layered structure. For all the structures investigated it is true that the electric field components parallel to the interfaces dominate in the skin and the muscle tissue, while in the fat layer the magnitude of the component normal to the interface is of the same order as the magnitudes of the parallel components. The width of the main lobes can be varied within limits by varying the media above the tissue.

The radiation properties of one dipole are briefly discussed in section 4.3.3. At the beginning of section 4.3.4, the ratio of $|\vec{E}|$ in the focus to the maximum $|\vec{E}|$ in the skin layer is introduced as a figure of merit for applicators. An applicator with a large figure of merit is preferred, because it reduces the danger of burning the skin layer. Considerations concerning the setup of the dipoles led to an array of 12 dipoles operating at 2.45 GHz.

As a first application a uniform current distribution of the dipoles is applied. The field distributions and the power absorbed per volume inside the tissue is given at great length for four different setups of the layers above the tissue. In the section 4.3.4.3 the general focusing condition is used, which guarantees a maximum field strength in the focus for a given input power. Comparing these results with those for uniform current distribution shows the limited value of the general focusing condition in the context of the problem considered here. Instead of this a procedure for optimizing all the parameters available is required. But as the analysis of only one setup is extremely time-consuming, such an optimization procedure has not been practicable till now. Thus only the most important parameter, the phase of the antenna currents, was systematically varied in order to specify optimal values. This led to an additional increase of the defined figure of merit by about 2 dB.

However, in any case, the maximum field strengths and, with this, the maximum absorbed power occurs inside the skin layer. On the other hand, in the highly lossy muscle tissue, the electric field, and thus also the temperature, decreases with increased distance from the interface; in other words, a temperature maximum is not achieved in the focus [4.59]. This situation is substantial improved if the skin surface is cooled with water [e.g. 4.16]. The two effects together, focusing and cooling, produce very promising results.

Finally the input impedance and the matching of the dipoles are investigated. As already shown in section 3.3.1, the input impedance of one single dipole is approximately equal to the input impedance of a dipole in homogeneous space with ε_{r0}. However, the mutual impedances between the dipoles are only slightly smaller than the self-impedances. Therefore, changing the mode of operation requires very different matching procedures.

It must be stressed that all considerations within chapter 4 refer only to the electromagnetic field induced inside the tissue and, by this, to the power absorbed per volume. The rise in temperature inside living tissue is not identical to the distribution of absorbed power, but is additionally determined by conductive and convective heat transfer phenomena due to blood circulation [4.60]. Therefore, after solving the electromagnetic boundary–value problem, the thermal problem must be considered. The separate treatment of these problems is possible, because the electrical properties of the biological material do not change significantly during heating within a range of some degrees. For the determination of the rise of temperature, the bio–heat equation has to be solved [4.61]. This problem turns out to be very complicated because it is nonlinear, but as a first–order representation linear heat transfer may be assumed. These problems will not be discussed in the context of this survey; the reader is referred to e.g.[4.61], which includes a comprehensive list of the most important papers.

CHAPTER 5

Geophysical probing with electromagnetic waves

5.1 Introduction

In the field of electromagnetic probing one usually attemps to estimate the depth, thickness, width and dielectric constant of subsurface scatterers from the geometrical and/or frequency variation of an electromagnetic field. Measurements are performed in the air, on the surface of the earth, in boreholes, in tunnels and caves. Frequencies are used from VLF up to the microwave region, the transmitter-receiver separations are in the range of some centimetres to some ten kilometres. The results are applied in geophysical prospecting, geotechnical engineering and communication system studies. Therefore it would be desirable if the required quantities could be determined directly from the measured data, i.e. if one could solve the inverse scattering problem. As is well known, inverse scattering problems are often intractable, and therefore one has to make maximum use of the information that can be obtained from a study of the direct scattering problem. (A review of inverse methods applicable to plane stratified media is given in [5.1].) What is needed, therefore, are catalogues of curves obtained by mathematical or scaled physical models. The measured data are then interpreted by matching observed and predicted data.

The first application of this method in connection with a stratified model of the earth to an electromagnetic field in the high frequency region is reported in papers by Großkopf [2.53] and [2.55], which have already been quoted in section 2.3.1. Großkopf used a two-layered model of the earth. He did not achieve complete agreement between model calculations and measured data in the frequency range considered (150 - 1500 kHz), because - as he noted himself - in this case it was not sufficient to use a model with only two layers and because the analytic treatment of the mathematical problem was not exact enough.

In the following decades up to the present day, most of the investigations which evaluate derived formulas numerically are restricted to models with maximally two layers. One of the reasons is that, as we have shown, the effort is considerable even with modern high-speed computers. Furthermore models with more layers contain so many parameters that a systematic discussion of the results is hardly feasible. Whereas Großkopf just as later on, for example, [5.2] to [5.8] used the wave tilt as the diagnostic tool, the majority of the work carried out up to now has been based on the induction method. This method is founded on the following considerations: A primary magnetic field is generated by an alternating current in a loop antenna or a grounded wire. If further conducting material is present within the magnetic field, currents are induced inside the conducting body. These eddy currents produce a magnetic field themselves, the secondary or induced field. The total field consisting of both parts, the primary and the secondary field, is measured for example with a second loop antenna. For a discussion of the results, the conducting body can be described by an equivalent circuit consisting of a resistance R and an inductance L in series. For sufficiently low frequencies the resistance R is dominating and the eddy currents flow deep within the conductor. With increasing frequencies the conductance L becomes important, and for high frequencies the skin effect restricts the current flow to a region near the surface of the body. In order to achieve deep penetration of the electromagnetic field into the earth crust, low frequencies have been used so far for the induction method, typically frequencies in the range of 100 to 5000 Hz. Therefore, the majority of the investigations using the induction method are concerned with low-frequency solutions neglecting displacement currents for antennas above stratified media. One of the very early papers is that of Wolf [2.51], who investigated the mutual coupling of two wires on a two-layer earth. His considerations are based on a formula for the mutual coupling of grounded wires, which had already been derived by Riordan and Sunde in 1933 for the application in the field of telephone communication [5.9]. At the beginning of the fifties Wait developed a theory for stratified media consisting of an arbitrary number of layers, as was outlined in section 2.3.1. In addition to the papers of Wait quoted there two more will be mentioned now: in 1951 he derived low-frequency formulas for a magnetic dipole [5.10] and in 1953 for a line source [5.11] over a stratified earth. In the following years up to today a series of investigations was carried out (for example [2.59],[2.60],[2.76],[2.77] and [2.115]) which was based mainly on the work of Wait. These investigations include those which were quoted in section 2.3.1 classified by the catchwords "quasi-stationary" and "quasi-static" (for example [2.63] to [2.69]).

The use of high frequencies leads to displacement currents which cannot be neglected compared to the conduction current and thus cause propagating waves inside and at the surface of the earth. In this sense the experiments of Großkopf in the frequency range of 150 to 1500 kHz can be considered as a first attempt of remote sensing of the ground with electromagnetic waves of high frequencies.

The method based on high-frequency propagation effects which has found the widest application so far is the radio interferometry technique. The essential features of this technique are illustrated in Fig. 5.1. A radio frequency source placed on the surface of the earth radiates energy both into the air above the earth and downward into the earth. Any subsurface contrast in electrical properties will result in some energy being reflected back to the surface. Thus, there will be interference maxima and minima in the field strength due to waves travelling different paths. The spatial positions of the maxima and minima are characteristic of the electrical properties of the earth and can be used as a method of inferring the geological properties of the earth.

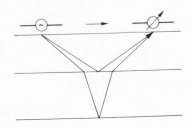

Fig. 5.1 Radio interferometry technique

Typically the earth's crust consists of dielectric material with high losses causing low depth of penetration of the waves radiated into earth and correspondingly low depth sounding. The radio interferometry technique therefore is especially apt for geologic regions with low electrical conductivity. The first reported application of the technique is the work of El-Said in 1956 [5.12], who attempted to sound the depth of the water table in the Egyptian desert. Because of the low conductivity of the geologic material above the water surface he successfully measured some interference maxima and minima, but as his interpretation of the data is only based on simple geometric-optical considerations the results are not very reliable.

It was not until the beginning of the seventies that a further series of investigations concerning the application of the radio interferometry technique followed. A principal motive was the experiment on electrical surface properties which was planned for the Apollo 17 mission. It was supposed that the lunar surface is very dry and therefore has a very low electrical conductivity. This fact promised a successful deep sounding. Extensive theoretical investigations in the context of this lunar experiment had been performed by Simmons et al. ([5.13] and [5.14]), Annan [5.15], Tsang et al. ([5.16] and [5.17]), Kong et al. [5.18]. In addition to this the papers of Fuller and Wait [5.19] and of Dey and Ward [5.20] should be mentioned here. As a preparation for the interpretation of lunar data the methods have been tested in the laboratory with analog models built to scale down and on glaciers (Rossiter et al.[5.21]), because ice has a low conductivity for radio frequencies.

Up to the present day the methods for the computation of the fields in or above layered structures have been improved more and more - this is especially the case for numerical analysis. The development is exemplified by the work quoted in section 2.3.1.

All papers quoted so far involve the analysis of the fields for one or more frequencies. Apart from this, the time domain response has been used in geophysical prospecting. As in the present chapter only investigations in the frequency domain will be discussed, we will not present the literature on transient electromagnetic methods here. The reader is referred to the comprehensive presentation of the development of transient methods in chapter 6.

Finally some monographs must be quoted. The books by Keller and Frischknecht [2.101], Wait [2.102] and King and Smith [2.103] have already been mentioned. Further important contributions are J.R. Wait: "Electromagnetic Probing in Geophysics" [5.22], Grant and West "Interpretation Theory in Applied Geophysics" [5.23] and Tsang, Kong and Shin: "Theory of Microwave Remote Sensing" [5.24].

5.2 Electrical properties of geological material

5.2.1 Some introductory remarks

The material of any i-th layer has so far been characterized by the permittivity $\varepsilon_0 \varepsilon_{ri}$ and the permeability $\mu_0 \mu_{ri}$. In order to be able to produce useful lists of curves the range of the values of ε_{ri} and μ_{ri} of materials of interest must be known. In the following, therefore, the electrical properties of the materials which are suitable for investigations with high-frequency fields - rocks, soil, ice and snow - will be briefly discussed.

For all materials mentioned the permeability is equal to that of free space, apart from a few exceptions. Therefore, we set $\mu_{ri} = 1$ so that only the complex dielectric constant $\varepsilon_{ri} = \varepsilon'_{ri} - j \varepsilon''_{ri}$ has to be discussed.

5.2.2 Rocks

A very detailed study of the electrical properties of rocks is given in the monograph of Parkhomenko [5.25], which also contains a comprehensive list of references. The following considerations are mainly based on this book.

Rocks are aggregates of numerous grains, bound together by molecular interaction forces. About 2000 minerals have been classified, but only a few tens of these minerals are common rock constituents. Feldspar is the most common rock-forming mineral group, comprising about 60% of the earth's crust (granite, syenite, basalt and so on). It is followed by quartz, which is the main constituent of quartzite and sandstone. The real part of the dielectric constant of feldspar is roughly 4.5 -6.5, of quartz roughly 4.7. For most of the other minerals which occur frequently, excluding the sulfides and oxides, the real part of the dielectric constant is in the range from 4 to 14. For some minerals, however, the values vary about two orders of magnitude from approximately 3 to 173 for rutile. It depends not only on the polarizability of the individual particles but also on the number per unit volume, which is closely related to the density, chemical composition and structure of a material.

Regarding the conductivity the materials of the earth can be classified as conductors, semiconductors or dielectrics. The highest conductivities are found in metals with values of conductivity ranging from 10^3 to 10^6 1/Ohm cm, whereas dielectrics have a conductivity of 10^{-9} 1/Ohm cm. Native metals occur infrequently, two of the more important native metals are copper and gold. The class of semiconductors includes the sulfide minerals, arsenides, graphite and a few oxides. Minerals of the third group with very high resistivities are widely distributed in the earth crust, for example quartz and calcite.

Rocks are very complicated materials, both in composition and structure. It is therefore not possible to give simple relationships between their electrical properties and their chemical or mineral composition. The common sedimentary and metamorphic rocks however are composed mostly of the minerals calcite, dolomite and quartz, yielding an ε'_r of about 5 - 9, where the true upper limit depends on whether a mineral with a high value of ε'_r is present as well.

The conductivity and so ε''_r of rocks covers a very wide range similar to that of minerals. Depending on which type of mineral is more important in a rock, rocks can be divided into two groups: rocks containing large amounts of low-resistivity minerals (for example ores) and rocks made up mainly of dielectrics. Models have been studied to calculate the conductivity of rocks consisting of grains distributed according to a conductive or resistive matrix. The grains are approximately described as spheres, ellipsoids, cubes etc.

All natural rocks contain some moisture with some degree of salinity. This has the effect that the electrical property is often dominated by the properties of the water, particularly in rocks with high water content such as sedimentary rocks. Even small amounts of moisture may result in a considerable change of the dielectric constant, which is explained by the fact that the conductivity of water is many orders of magnitude higher than the conductivity of most rock-forming minerals. The exact nature of the relationship between bulk resistance and water content depends on the type of rock, the porosity and the salinity of the water.

In any case the water content is responsible for a strong dependence on frequency as was the case for biological material (chapter 4). As an example, results reported by Klein [5.26] for the complex dielectric constant of coal as a function of frequency for a water content of 0.8%, 10.2% and 29.3%, are quoted. In the first case the imaginary part is mainly determined by the dc conductivity of coal; it varies from about 0.45 for 100 Hz to about 0.03 for 1 GHz. The real part decreases from 3.2 to 2.0. A medium water content leads to an increased dc conductivity and thus to increased values for ε''_r for low frequencies. In the range 1 - 100 MHz the Maxwell-Wagner-relaxation is observed. This effect is no longer relevant for a water content of 29.3%, where up to about 2 GHz ε''_r is dominated by the ionic conductivity, and for even higher frequencies by the relaxation of free water.

Most minerals are crystalline and therefore exhibit anisotropy. This often results in an anisotropy of a rock, too. This is true as well if the inclusions are elongate or have needle form. In both cases the dielectric constant must be described by an ε_r-tensor.

5.2.3 Soil

A soil medium is a four-component dielectric mixture consisting of air, bulk soil, bound and free water [5.27]. The bulk material is classified as sand for diameters $d > 0.05$ mm, as silt for $0.002\text{mm} \leq d \leq 0.05$ mm and as clay for $d < 0.002$ mm. The solid ingredients of the soil have various shapes, a distinctive particle size distribution and a characteristic specific density ρ_s. They are randomly packed into a structure with a bulk density ρ_b. The clay mineral fraction can exist both as discrete particles and as coatings on coarser particles. The porosity V_p is given by $V_p = 1 - \rho_b/\rho_s$, the pores are filled with water and/or air. The amount of bound water mainly depends on the entire soil particle surface and on the thickness of the bound water layer, typically some Å. Formulas for the calculation of the volume fraction of bound water and bulk water are given in [5.28]. The dielectric properties of bound water have not been known until now. Experimental results indicate that in the microwave region both the real and imaginary parts are considerably lower than for free water and that bound water exhibits a dielectric relaxation lower than that of pure water. In any case, both ε'_r and ε''_r increase strongly with increased moisture content, because of the much higher dielectric constant of water than that of the dry soil components.

The dielectric constant of soil also depends on the soil texture. For the soils investigated in [5.28], ε'_r was found to be roughly proportional to sand content and inversely proportional to clay content. As for ε''_r, for low microwave frequencies the effective ionic conductivity of the soil is dominant, for high frequencies the dielectric relaxation is the principal mechanism. This results in an increase of ε''_r with soil clay content for low frequencies, a nearly constant ε''_r for all soil textures in the frequency range from about 4.0 to 6.0 GHz and a decrease of ε''_r with soil clay fraction for higher frequencies.

Experimental data for soils are reported for example in [5.27] to [5.32] for the UHF and microwave region. For dry soil the values are lower than those for dry rocks (for example $\varepsilon'_r \approx 2.5...4.5$), for high water content ε'_r increases up to about 30 and ε''_r up to about 15.

Several dielectric models have been developed in order to derive equations for the dielectric constant on the basis of physical soil parameters ([5.28] and [5.31] to [5.34]). Good results are obtained with the equation given in [5.28], which is based on the mixture formula of de Loor [5.35]. For that the dielectric constant of the dry soil material is determined from experimental data from [5.27] and that of the free water is described by the Debye-type relaxation modified to account for the ionic conductivity losses. The values for the dielectric constant of bound water are estimated from the experimental results.

5.2.4 Ice

The complex dielectric constant of pure ice can be calculated as a function of temperature and frequency by the simple Debye dispersion formula

$$\varepsilon_{r\,ice} = \varepsilon'_{\infty ice} + \frac{\varepsilon'_{r0\,ice} - \varepsilon'_{\infty ice}}{1 + \omega\tau_{ice}} \tag{5.1}$$

with

$$\varepsilon'_{\infty ice} = 3.14 \tag{5.2}$$

$$\varepsilon'_{r0\,ice} = 20715/(T + 235) \tag{5.3}$$

$$\tau_{ice} = 5.3 \cdot 10^{-6} e^{756.4/(T + 273)}{}_{sec} \tag{5.4}$$

where the temperature T is given in °C [5.36]. According to this formula the real part of the dielectric constant drops from a low-frequency limiting value of about 100 to a high-frequency value of 3.14. The dc conductivity can be neglected for high-frequency considerations. The loss tangent obtained at 10 GHz is of the order of 10^{-3}. Detailed experimental data for pure ice are reported for example in [5.37] to [5.40] for the microwave region.

When sea water freezes, however, the resulting ice contains entrapped brine. Therefore, the electrical properties of sea ice which can be found in nature are quite different from those of pure ice. According to [5.41], the major mechanisms at high frequencies are the Maxwell-Wagner-Sillars dispersion, for frequencies in the MHz region Debye dispersion of the ice protons with reduced relaxation time and for low frequencies dispersion resulting from ion-space charge polarization in the brine channels. For the UHF and microwave frequencies the model derived in [5.42] and [5.43] indicates that ε''_r of sea ice is directly proportional to the volume fraction of brine in the ice V_{br} and that ε''_r is approximately given by

$$\varepsilon'_r = a_0 + \frac{a_1}{1 - 3V_{br}} \tag{5.5}$$

where $a_{0/1}$ are determined from experimental data. According to [5.36] the real part of the dielectric constant lies in the range $3.5 \leq \varepsilon'_r \leq 5.0$ for salinities of the ice up to 1.0 % and frequencies from 0.1 GHz to 40 GHz. The upper limit is valid for about −4° C, the values decrease with decreasing temperature. The corresponding values for ε''_r lie between 0.03 and 1.0, again the lower values belong to lower temperatures.

Summarizing, it can be stated that pure ice (and ice from fresh water of course) is a penetrable medium and sea ice an absorptive medium for electromagnetic waves in the UHF and microwave regions.

5.2.5 Snow

Snow is a mixture of three components: ice particles, liquid water and air. As the complex dielectric constants of ice and water depend on frequency and temperature, so too does the dielectric constant of snow. In addition it is a function of snow density, ice-particle shape and the shape of water inclusions.

In the earlier literature there are only few data available for the dielectric constants of snow in the high-frequency and microwave regions ([5.37] and [5.44] to [5.50]). Recently a comprehensive study has been published by Hallikainen et al. [5.51] for the range of 3 to 18 GHz and for 37 GHz. A distinction is made between dry and wet snow for the discussion of the data. Dry snow is a mixture of ice and air only. Just as the real part of the dielectric constant of ice is independent of temperature and frequency in the microwave region, so too is it for the dielectric constant of dry snow. The dielectric constant then depends only on the snow density and can be calculated by the mixing formula of Polder and Van Santen [5.52]. For the real part, values from 1.0 to about 3.0 are obtained. The loss tangent of dry snow is less than that of ice, of course, which means that it is very low (≈ 0.001 at 1 GHz). Therefore, microwaves can deeply penetrate layers of dry snow.

Wet snow consists of dry snow and water inclusions. The real part of the dielectric constant of water is about 40 times larger than that of dry snow. Hence, even when the liquid water content is of the order of only 1% by volume, the spectral behaviour of wet snow is dominated by the dispersion behaviour of water. It further depends on the shape of the water inclusions, which may vary as a function of water content. For the calculation of the dielectric constant, several dielectric mixing models have been considered in the literature (for example Polder and Van Santen models, Debye-like models [5.51]), based on different shapes of the water inclusions (for example randomly oriented ellipsoids) or on empirical data.

According to results reported in [5.51] the real part of the dielectric constant varies from about 2 to 6 and the imaginary part from 0.001 to 1 in the microwave region.

5.3 Model calculations

5.3.1 Basic considerations

As a result of the previous section we can state that the real part of the dielectric constant of most of the geological material of interest is in the range from about 1.0 to about 14.0. Therefore, ε'_r is varied within this range for the models discussed in this section. The imaginary part of ε''_r covers the very wide range from low-loss ($\varepsilon''_r \lesssim 0.01$) to highly lossy material ($\varepsilon''_r \gtrsim 10.0$). The results shown below confirm the assumption that it is worthwhile to consider only the range $0.01 \lesssim \varepsilon''_r \lesssim 10.0$.

The results obtained for our model can be transferred to other models using the rules stated for example in [5.53]. This approach is often very advantageous because in this way large geological fields can be simulated by small scale models of a size which can be handled in a laboratory. According to [5.53] qualitative models can be used, which directly yield results of those properties not dependent on power level (impedance, polarization, relative pattern), or quantitative models, which are capable of giving quantitative data on all electromagnetic properties of the system, for example field intensity. The reduction in size leads, for example, to an increase in conductivity and of frequency or a combination of these factors.

The starting point of all the theoretical considerations in this book is the model of a plane stratified medium, which extends to infinity. Therefore, it has to be checked to what extent this kind of model is applicable in geology.

All methods for the investigation of the structure of the earth with electromagnetic waves depend on the measurement of the field of an antenna. An inhomogeneity in the ground to be investigated can only be detected if the level of the field strength is sufficiently high at the position of the inhomogeneity. Vice versa, inhomogeneities in areas of sufficiently low field strengths have no influence on the measurement. These

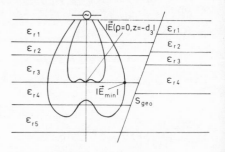

Fig. 5.2 Layered structure with a fault S_{geo}

circumstances are shown schematically in Fig. 5.2: The interface to be measured is supposed to be located at a depth of $z = -d_3$, the level which the antenna generates at the position of the interface is $|\vec{E}(\rho = 0, z = -d_3)|$. The region which must be free of faults, which are not described by the plane layer model, is bounded by the line $|\vec{E}_{min}|$, where the difference of levels $\Delta|\vec{E}| = |\vec{E}(\rho = 0, z = -d_3)| - |\vec{E}|_{min}$ is given by the required accuracy and the apparatus used. The fault S_{geo} which is sketched in the figure consequently

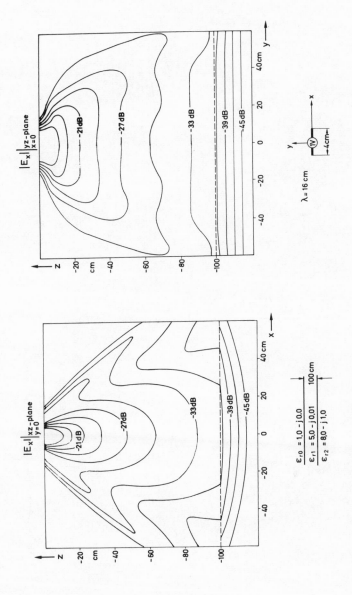

Fig. 5.3 Electrical field of a dipole on a two-layered earth

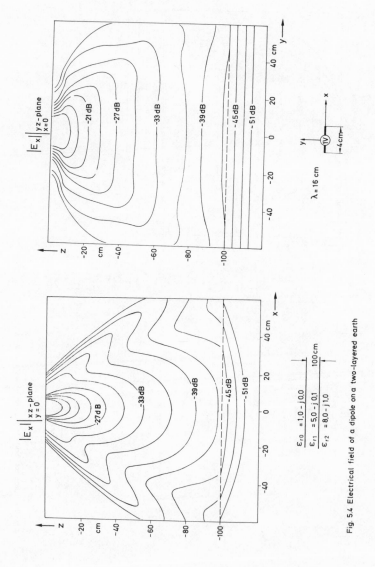

Fig. 5.4 Electrical field of a dipole on a two-layered earth

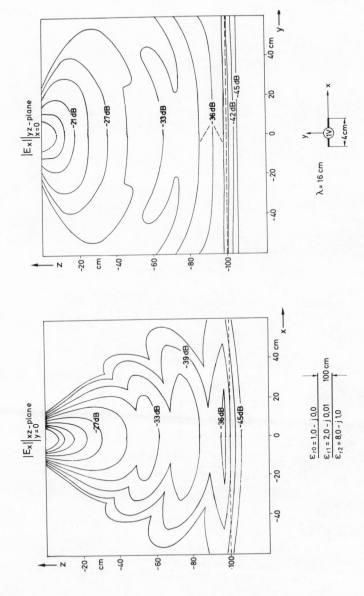

Fig.5.5 Electrical field of a dipole on a two-layered earth

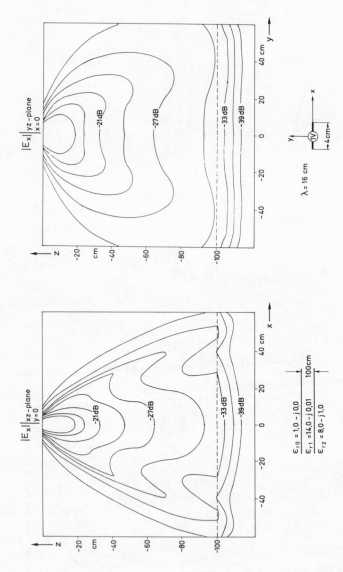

Fig.5.6   Electrical field of a dipole on a two-layered earth

would not prevent a successful measurement of the interface $z = -d_3$. In a practical application one can determine the region which must be free of faults by calculating the absolute value of the field strength in the earth. Figs. 5.3 to 5.6 show examples of the level of the field strength for four different structures consisting of two layers. The field is generated by a dipole of length 4 cm at a frequency of 1.875 GHz ($\lambda = 16$ cm). The dipole operates just below the resonance frequency (The length of resonance is, with $\varepsilon'_{r1} = 5$, $\varepsilon''_{r1} = 0$, equal to 4.62 cm). The data for the field strength are normalized by the absolute value of the field strength at the position $\rho = 0, z = -1$ cm. (The fixing of the normalization is largely arbitrary as only the relative values are relevant for the following discussion). In all cases $\varepsilon_{r2} = 8.0 - j1.0$. Figs. 5.3 and 5.4 show the results for $\varepsilon'_{r1} = 5.0$ and $\varepsilon''_{r1} = 0.01$ and 0.1 respectively. Just as for fields in biological tissue (cf. chapter 4) only the x-component of the E-field has to be taken into account, if the dipole antenna is oriented in the x-direction. Due to the higher attenuation of the structure of Fig. 5.4 the level $|\vec{E}(\rho = 0, z = -d_1)|$ is lower by 6.5 dB for this example compared to the structure of Fig. 5.3. This value corresponds roughly to the higher attenuation of a plane uniform wave in a homogeneous medium with $10 \cdot \varepsilon''_{r1}$ compared to a medium with ε''_{r1}. As the dipoles point into the x-direction the focusing of the radiation fields in the xz-plane is slightly stronger than in the yz-plane. As was to be expected, in both cases the lateral extension of the fields is very large. For example, for the structure of Fig. 5.3 one obtains that the level diagram -36 dB $\approx |\vec{E}(\rho = 0, z = -d_1)|$ extends in the yz-plane in the region $|y| \lesssim 100$ cm (in the figures only the regions $|x| \leq 60$ cm, $|y| \leq 60$ cm respectively have been presented).

Analogously to the procedure of chapter 4 the extension of the field will be marked by the 3-dB-width in the plane $z = -d_1$ (cf. eqns. (4.11)- (4.15), (4.17)).

One obtains for the structure of

Fig. 5.3 $b_{gx}|_{xz-\text{plane}} \approx 40$ cm, $b_{gx}|_{yz-\text{plane}} \approx 130$ cm

and

Fig. 5.4 $b_{gx}|_{xz-\text{plane}} \approx 36$ cm, $b_{gx}|_{yz-\text{plane}} \approx 90$ cm.

   Figs. 5.5 and 5.6 show the level diagrams for $\varepsilon_{r1} = 2.0 - j0.01$ and $\varepsilon_{r1} = 14.0 - j0.01$. Because of the strongly differing values of ε_{r1} and ε_{r2} in Fig. 5.5 strong reflections occur at the interface. They lead to standing waves in the region of layer 1. The 3-dB-values are for the structure of

Fig. 5.5 $b_{gx}|_{xz-\text{plane}} \approx 45$ cm, $b_{gx}|_{yz-\text{plane}} \approx 150$ cm

and

Fig. 5.6 $b_{gx}|_{xz-\text{plane}} \approx 55$ cm, $b_{gx}|_{yz-\text{plane}} \approx 65$ cm.

Note that for all examples shown the field strength decreases rapidly in the region of the surface of the earth. Faults in this area consequently do not have so much effect as faults near the interface to be measured.

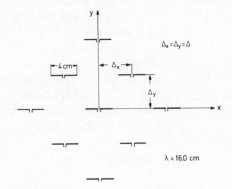

Fig. 5.7   Array of 9 dipoles

5.3.2 Improvement of the resolution using arrays

Narrower mainlobes are obtained by the use of arrays of antennas. At first sight it seems obvious that focusing antennas like the ones discussed in chapter 4 should be employed. For focusing antennas, however, the diameter of the array is typically at least double as large as the depth of the layer to be measured (in chapter 4 the diameter of the antenna is 9 cm, the focus is located at a depth of 4 cm). This implies a very large region which has to be free of faults. It would be more advantageous to use smaller arrays of antennas such that the region to be measured lies in the farfield of the array. Figs. 5.8 and 5.9 again show some examples. The data for the layers are the same as for the structures of Fig. 5.3. The setup of the 9 dipoles (length 4 cm) is sketched in Fig. 5.7. Only results for the xz-plane are given as those for the yz-plane are very similar. The reason is that the field pattern of the array is principally determined by the characteristic of the group and not so much by the characteristic of the single antenna. The results show that - just as for antennas in homogeneous space - the width of the mainlobe decreases with growing distances $\Delta x, \Delta y$. For example, for $\Delta x = \Delta y = 4$ cm one obtains a 3-dB - width of 22 cm, for $\Delta x = \Delta y = 6$ cm the result is b_{gx}=11 cm and for $\Delta x = \Delta y = 10$ cm the value for b_{gx} is only 7 cm. On the other hand the level of the sidelobes grows with increasing distances. For $\Delta x = \Delta y = 4$ cm the line $|\vec{E}(\rho = 0, z = -d_1| - 10dB = -27dB$ only shows a weakly pronounced sidelobe, for $\Delta x = \Delta y = 10$ cm, however, the field strengths in the points $z = -d_1, x = 0$ (mainlobe) and $z \approx -50$ cm, $x \approx 50$ cm (second sidelobe) are equal. The reason for the appearance of especially large sidelobes

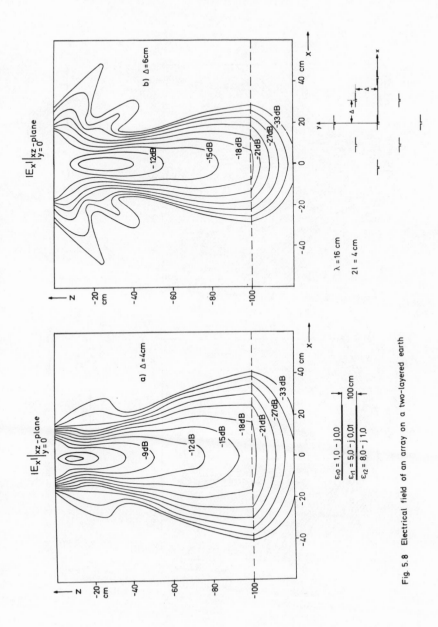

Fig. 5.8 Electrical field of an array on a two-layered earth

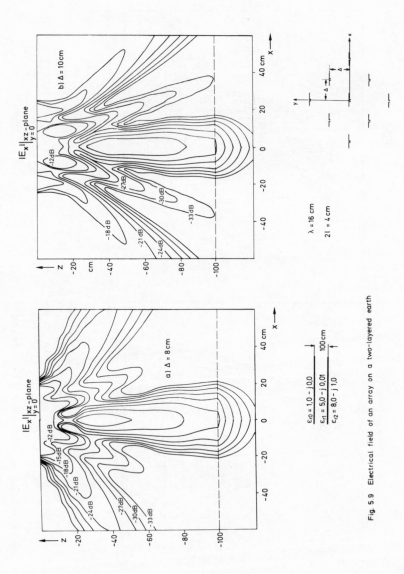

Fig. 5.9 Electrical field of an array on a two-layered earth

in the latter case is that the distances of the radiation elements are greater than $\lambda/\sqrt{\varepsilon'_{r1}} = 7.2$ cm which means that grating lobes appear.

A narrower mainlobe avoiding growing sidelobes is obtained if more radiation elements are used, the distance of which must be smaller than $\lambda/\sqrt{\varepsilon'_{r1}}$. Note, however, that for many radiation elements the expenditure of computer time and storage capacity increases rapidly.

5.3.3 Mutual coupling of two dipoles above a multilayered earth

This section introduces some model calculations for the mutual coupling of two dipole antennas over stratified structures.

First of all we will discuss how a suitable setup of dipoles on the earth can be found for a given measurement task. To this end we imagine that the depth d_1 and the dielectric constant ε_{r2} of the stratified structures investigated in the last section have to be measured. From the figures of the field pattern in the earth (Figs. 5.3 to 5.6) it can be inferred that the distance of the centres of the dipoles should be roughly 20 cm. A considerably smaller distance leads to a strong direct coupling of the two dipoles, whereas for a larger distance the receiving dipole lies in a region of too low field strength. As the field generated by a dipole oriented in the x-direction essentially only has an E_x-component, the dipoles are oriented parallel to each other (Fig. 5.10).

For a fixed frequency a favourable location of the receiving dipole with respect to the transmitting dipole is determined by calculating the mutual coupling for different $\Delta x, \Delta y$ for the case of $d_1 \to \infty$, i.e. without the second layer. The sensitivity to an interface at a finite depth d_1 is especially large, if $\Delta x, \Delta y$ are chosen such that the two dipoles are decoupled for $d_1 \to \infty$. The numerical effort, however, needed for the determination of the optimal position is considerable, as first of all Green's function for the entire region $0 < \rho \lesssim 22$ cm has to be

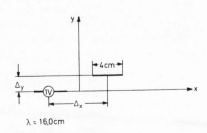

Fig. 5.10 Position of two dipoles over a stratified earth

calculated and then the mutual coupling for every possible combination $\Delta x, \Delta y$ has to be evaluated. In order to reduce the effort, Green's function was only determined for the regions $0 < \rho < 2l = 4$ cm and 14 cm $< \rho < 22$ cm. The mutual coupling was calculated only for positions given in the table below, which shows as an example the mutual coupling of two dipole antennas, coupled via a homogeneous half space with $\varepsilon_{r1} = 5.0 - j0.01$. Obviously $|Z_{21}|$ is minimal for position 5. If for a certain structure a distinct minimum for $|Z_{21}|$ could not be found, we chose that position for which $|X_{21}|$ was minimal so that the interface at $z = -d_1$ was noticeable in the largest possible phase shift of Z_{21}.

Case	Δx	Δy	ReZ_{21}/Ω	ImZ_{21}/Ω	$|Z_{21}|/\Omega$
1	180 mm	0 mm	-0.196	0.068	0.208
2	175 mm	20 mm	-0.200	0.088	0.218
3	175 mm	40 mm	-0.154	0.050	0.162
4	170 mm	60 mm	-0.103	0.014	0.104
5	**160 mm**	**80 mm**	**-0.023**	**-0.012**	**0.026**
6	150 mm	100 mm	0.063	-0.092	0.111
7	140 mm	120 mm	0.087	-0.240	0.255
8	120 mm	140 mm	0.189	-0.365	0.411
9	110 mm	160 mm	-0.130	-0.513	0.529
10	80 mm	180 mm	-0.249	-0.621	0.669
11	40 mm	200 mm	-0.544	-0.495	0.763

As a first example Figs. 5.11 to 5.14 show the mutual coupling on two-layer structures for $\varepsilon'_{r1} = 2.0, 5.0, 8.0$ and 14.0, $\varepsilon''_{r1} = 0.01$ as a function of ε'_{r2} ($\varepsilon''_{r2} = 1.0$ and $\varepsilon''_{r2} = 10$). The thickness of the layer is $d_1 = 100$ cm, the frequency is 1.875 GHz ($\lambda = 16$ cm). The optimal location of the dipoles with respect to each other is specified by Δx and Δy. The absolute values of the coupling impedances are in the range $0 < |Z_{21}| \lesssim 1.0\Omega$. As the input impedances Z_{11} and Z_{22} of the two dipoles vary only slightly with ε_{r2} and $|Z_{11}| = |Z_{22}| \gtrsim 20\Omega$ holds for the absolute values, the mutual impedances can easily be converted to mutual admittances because of

$$Y_{21} = \frac{Z_{21}}{Z_{11} \cdot Z_{22} - Z_{12} \cdot Z_{21}} \approx \frac{Z_{21}}{Z_{11}^2} \tag{5.6}$$

In order to simplify the interpretation of the curves the reflection coefficient

$$r = \frac{\sqrt{\varepsilon_{r1}} - \sqrt{\varepsilon_{r2}}}{\sqrt{\varepsilon_{r1}} + \sqrt{\varepsilon_{r2}}} = |r|e^{j\delta} \tag{5.7}$$

for vertical incidence of a homogeneous plane wave on the interface $z = -d_1$ is shown as well. For $\varepsilon_{r1} = (2.0 - j0.01)$ and $\varepsilon''_{r2} = 1.0$ (Fig. 5.11) $|r|$ grows continuously with ε'_{r2}. This leads to a strong increase of R_{21}. As the reflection coefficient for $\varepsilon''_{r2} = 10.0$ does not vary appreciably with ε'_{r2}, the variation of Z_{21} is small as well. Figs. 5.12 and 5.13 are marked by the fact that in the region considered the absolute values of the reflection coefficients for $\varepsilon'_{r1} \approx \varepsilon'_{r2}$ and $\varepsilon''_{r2} \approx 1.0$ show a distinct minimum. The comparison of all curves in Figs. 5.11 to 5.14 shows that the behaviour of R_{21} differs considerably from that of X_{21}. Further calculations have shown that this behaviour and thus the sensitivity for

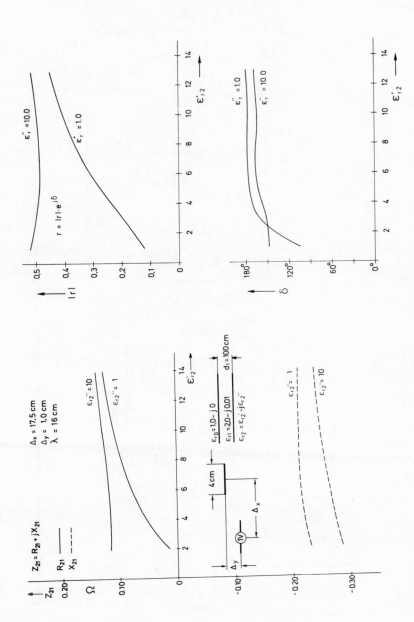

Fig. 5.11 Mutual impedance of two dipoles and reflection coefficient for vertical plane wave incidence at $z = -d_1$ as a function of ε'_{r2}

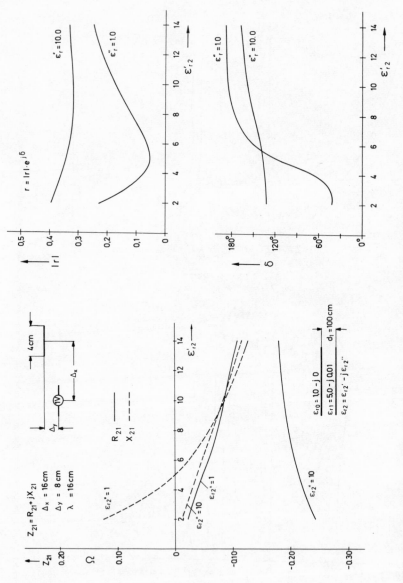

Fig. 5.12 Mutual impedance of two dipoles and reflection coefficient for vertical plane wave incidence at $z = -d_1$ as a function of ε'_{r2}

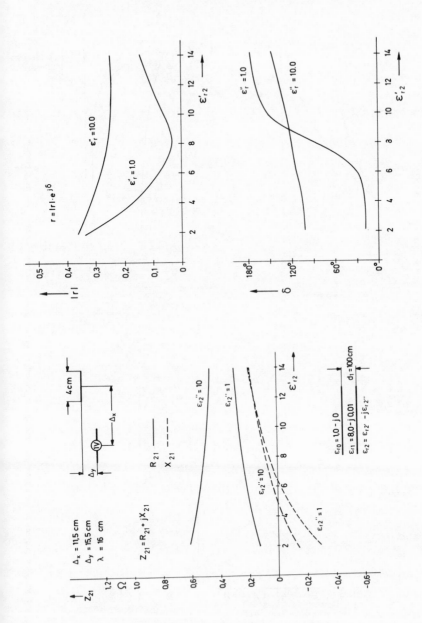

Fig. 5.13 Mutual impedance of two dipoles and reflection coefficient for vertical plane wave incidence at $z=-d_1$ as a function of ε''_{r2}

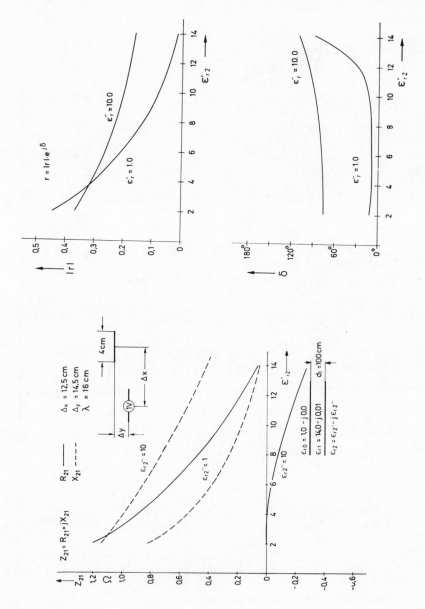

Fig. 5.14 Mutual impedance of two dipoles and reflection coefficient for vertical plane wave incidence at $z = -d_1$ as a function of ϵ'_{r2}

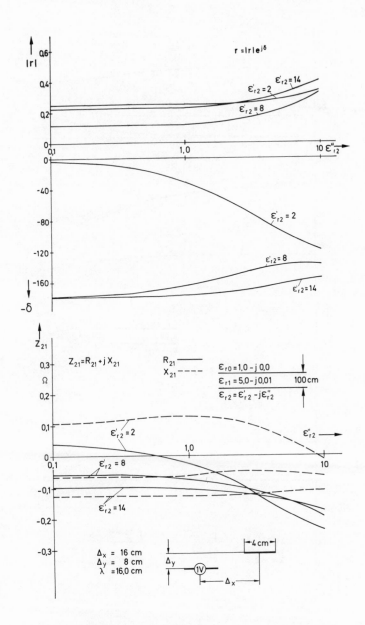

Fig. 5.15 Reflection coefficient for vertical plane wave incidence at $z = -d_1$
and mutual impedance of two dipoles as a function of ε''_{r2}

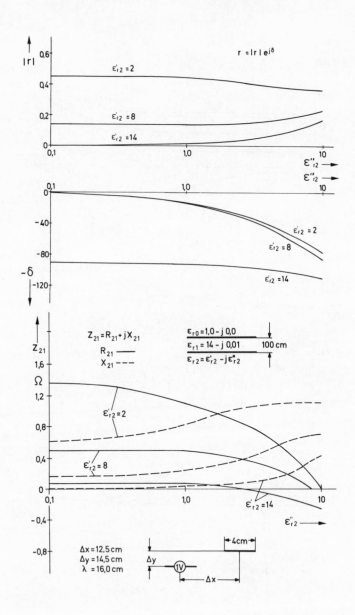

Fig. 5.16 Reflection coefficient for vertical plane wave incidence at $z = -d_1$ and mutual impedance of two dipoles as a function of ε''_{r2}

the measurement task under consideration is strongly dependent on the location of the dipoles with respect to each other.

Figs. 5.15 and 5.16 show the mutual impedances as functions of ε_{r2}'' for the structures of Figs. 5.12 and 5.14. As the reflection coefficient for $\varepsilon_{r2}'' \lesssim 0.1$ is nearly constant, the amount of losses of low-loss material in layer 2 cannot be measured.

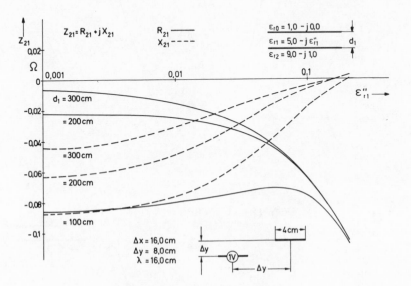

Fig. 5.17 Mutual impedance of two dipoles as a function of ε_{r1}''

The penetration depth of electromagnetic waves, and thus the depth up to which the interface $z = -d_1$ can be measured, is substantially determined by the losses in layer 1. In Fig. 5.17 Z_{21} is plotted as a function of ε_{r1}'' for $d_1 = 100, 200$ and 300 cm. For $\varepsilon_{r1}'' \gtrsim 0.2$ all curves meet so that a measurement of d_1 is no longer possible. The curves confirm the statement made earlier that methods based on electromagnetic waves cannot be applied successfully if layers with high losses have to be penetrated.

The calculation of Z_{21} as a function of d_1 with fixed material parameters (Figs. 5.18 and 5.19) yield results which are easy to interpret. R_{21} and X_{21} oscillate round a mean value which is equal to the mutual impedance for $d_1 \to \infty$. The period is equal to

$$\lambda/2\sqrt{\varepsilon_{r1}'} \;\; ; \varepsilon_{r1}' \gg \varepsilon_{r1}'' \quad , \tag{5.8}$$

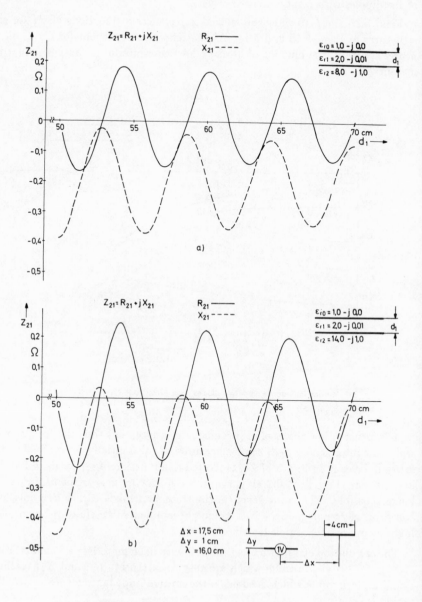

Fig. 5.18 Mutual coupling of two dipoles as a function of d_1

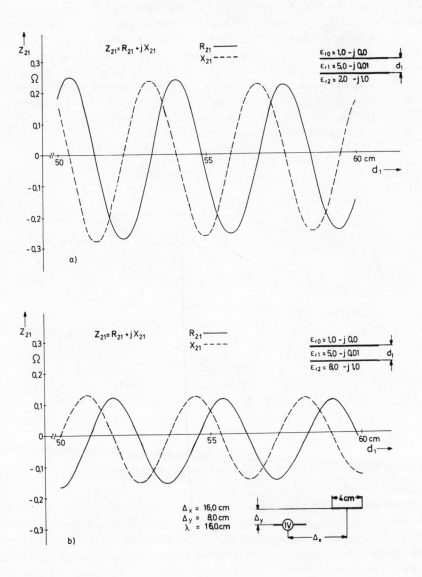

Fig. 5.19 Mutual impedance of two dipoles as a function of d_1

The curves of R_{21} and X_{21} are shifted by a quarter of a period with respect to each other. In Fig. 5.18b the amplitude of the oscillations is larger compared to Fig. 5.18a, as in the first case $|r|$ is larger. The same holds for Figs. 5.19a and 5.19b. Due to the damping in layer 1 the amplitudes of the oscillations in all cases decrease with growing d_1. The corresponding curves of Figs. 5.18a and 5.18b are nearly equal in phase, as the phase angles of the reflection coefficients differ only by approximately $3°$. In the case of Figs. 5.19a and 5.19b the difference between the phase angles of r is about $-140°$. This corresponds to a shift of 1.36 cm of the curves with respect to each other.

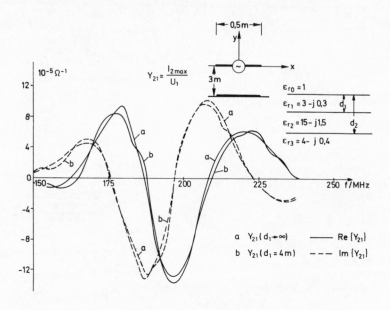

Fig. 5.20 Mutual impedance of two parallel dipoles as a function of frequency

For a structure with many layers a systematic variation and discussion of all relevant parameters is not feasible. In such a case the investigation must be guided by the actual measurement task under consideration in order to obtain the optimal dimensioning and setup of the dipoles. In the following, therefore, only a three-layer structure and a four-layer structure will be discussed, whereby now the frequency dependence is investigated. In this case one cannot use optimal values for Δx and Δy as these depend on frequency. For that reason one has to start off with a fixed setup of the dipoles. Frequently the influence of single interfaces can be determined directly from the results. As an example, in Fig. 5.20 the coupling admittance of two parallel dipole antennas is displayed in the frequency range of 150 - 250 MHz. The curves a) have been calculated for $d_1 \to \infty$, i.e. for the homogeneous half-space with $\varepsilon_{r1} = 3 - j0.3$, the curves b) for $d_1 = 4$ m. The curves are very similar, as the coupling impedance of the two dipoles is essentially determined by the dipole length, the distance and ε_{r1}. One obtains the influence of layers 2 and 3 together by calculating $|Y_{21}(d_1 = 5 \text{ m})|/|Y_{21}(d_1 \to \infty)|$ (see curve a) in Fig. 5.21).

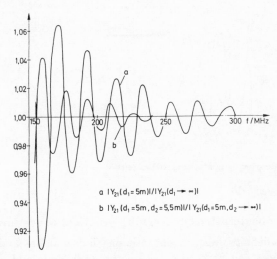

Fig. 5.21   Normalized mutual coupling of two parallel
dipoles (for parameters of the structure see Figure 5.20)

For sufficiently high frequencies ($f \gtrsim 250$ MHz) curve a) is exclusively determined by the interface of $z = -d_1$ because of the high attenuation in layer 2. The influence of the interface $z = -d_2$ becomes visible if the ratio $Y_{21}(d_1 = 5 \text{ m}, d_2 = 5.5 \text{ m})/Y_{21}(d_1 = 5 \text{ m}, d_2 \to \infty)$ is plotted (curve b) in Fig. 5.21). If the layers are sufficiently lossy the single layers can easily be investigated by measurements with varying frequencies; lower and lower layers are reached by decreasing the frequency.

For thin and low-loss layers the contributions from the single layers cannot be resolved that easily. As an example, Fig. 5.22 shows the results for a 4-layer structure. The determination of the layer parameters now requires the application of lengthy procedures for the solution of the inverse scattering problem. With the help of the model calculations that have been introduced here, however, one can fix suitable setups of the antennas and frequency ranges beforehand, and the influence of faults, which are not accounted for by the plane stratified layer model, can be estimated.

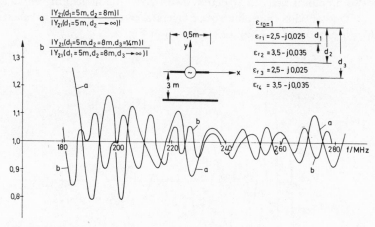

Fig. 5.22   Normalized   mutual   coupling   of two   parallel   dipoles

5.3.4 Some considerations concerning practical procedure

If one has a definite measurement task, one first of all has to decide whether the investigations can be carried out on the given scale and with the given materials, or whether a scale– and/or material–transformation has to be made before. If the scale factors for the electric field strength $\vec{E}$, the magnetic field strength $\vec{H}$, the time t and the position vector $\vec{r}$ are defined by

$$\vec{E}(\vec{r},t) = \alpha_e \vec{E}_{tr}(\vec{r}_{tr}, t_{tr}) , \tag{5.9}$$

$$\vec{H}(\vec{r},t) = \alpha_h \vec{H}_{tr}(\vec{r}_{tr}, t_{tr}) , \tag{5.10}$$

$$t = \alpha_t t_{tr} , \tag{5.11}$$

$$\vec{r} = \alpha_r \vec{r}_{tr} , \tag{5.12}$$

then, according to [5.53], the electrical properties of the model systems can be given in terms of the properties of the full-scale systems and the scale factors α_e, α_h, α_t and α_r by:

$$\varepsilon_{tr} = \frac{\alpha_r \alpha_e}{\alpha_h \alpha_t} \varepsilon \tag{5.13}$$

and

$$\mu_{tr} = \frac{\alpha_r \alpha_h}{\alpha_e \alpha_t} \mu \ . \tag{5.14}$$

Note that for the conductivity κ

$$\kappa_{tr} = \frac{\alpha_r \alpha_e}{\alpha_h} \kappa \tag{5.15}$$

holds. As for most natural materials and also for materials which are available for the models

$$\mu \approx \mu_0 \tag{5.16}$$

holds, one can set

$$\frac{\alpha_r \alpha_h}{\alpha_e \alpha_t} = 1 \ . \tag{5.17}$$

Furthermore it is often appropriate to leave the space above the layers unchanged, i.e.

$$\varepsilon_{r0} = \varepsilon_{r0\,tr} = 1 \ . \tag{5.18}$$

Thus

$$\alpha_e = \alpha_h \tag{5.19}$$

and

$$\alpha_r = \alpha_t \tag{5.20}$$

or

$$f_{tr} = \alpha_t f \ , \tag{5.21}$$
$$\varepsilon_{tr} = \varepsilon \tag{5.22}$$

and

$$\kappa_{tr} = \alpha_r \cdot \kappa \ . \tag{5.23}$$

For geological model investigations in the laboratory often very large values for α_r have to be used ($\alpha_r = 10^2 \ldots 10^5$). A substantial problem consequently is that materials with a very high conductivity are needed. Suitable materials are metals and for conductances in the region of $10^4 1/\Omega m \ldots 10^5 1/\Omega m$ carbon, graphite (unloaded or loaded) or epoxy (copper or silver loaded).

The choice of the dimensions of the model has to be seen in close connection with the establishment of the frequency. Strong reductions simultaneously imply high frequencies. On the other hand the frequency which should be used is determined by the required penetration depth. As has been discussed in section 5.2, earth material usually is lossy and thus effects a strong attenuation. For layers which are not too thin, one obtains a good assessment of the attenuation as a function of the frequency by considering a plane homogeneous wave, which penetrates the earth vertically, and merely taking into account the contributions due to reflections at the single layer boundaries but not the multiple reflections: The power flow density just above the interface $z = -d_I$ then is a function of the power flow density at $z = 0$

$$S_e(z=-d_I) \approx S_e(z=0) \cdot \left|\frac{Z_I}{Z_0}\right|^2 \frac{Re\{Z_I\}}{Re\{Z_0\}} \prod_{i=1}^{I} |1+r_{i-1}|^2 \; e^{-2\alpha_i(d_i-d_{i-1})} \quad (5.24)$$

with

$$\alpha_i = -Im\{\omega\sqrt{\mu_0\varepsilon_0} \cdot \sqrt{\varepsilon'_{ri} - j\varepsilon''_{ri}}\} \quad (5.25)$$

and

$$r_i = \frac{\sqrt{\varepsilon_{ri}} - \sqrt{\varepsilon_{ri+1}}}{\sqrt{\varepsilon_{ri}} + \sqrt{\varepsilon_{ri+1}}} \; ; \quad (5.26)$$

$$Z_i = \sqrt{\frac{\mu_0}{\varepsilon_0}} \sqrt{\frac{1}{\varepsilon_{ri}}} \; . \quad (5.27)$$

The contribution which is reflected at the interface $z = -d_I$ and propagates to the position $z = 0$ hence is

$$S_{rI}(z=0) =$$

$$= S_e(z=0)|r_I|^2 \cdot \prod_{i=1}^{I} \left[|1+r_{i-1}|^2|1-r_{i-1}|^2 e^{-4\alpha_i(d_i-d_{i-1})}\right] \quad (5.28)$$

$$= S_e(z=0)\left|\frac{\sqrt{\varepsilon_{rI}} - \sqrt{\varepsilon_{rI+1}}}{\sqrt{\varepsilon_{rI}} + \sqrt{\varepsilon_{rI+1}}}\right|^2 \prod_{i=1}^{I} \left[\left|\frac{4\sqrt{\varepsilon_{ri}}\sqrt{\varepsilon_{ri+1}}}{\sqrt{\varepsilon_{ri}} + \sqrt{\varepsilon_{ri+1}}}\right|^2 e^{-4\alpha_i(d_i-d_{i-1})}\right] .$$

For the example shown in Fig. 5.20 this equation yields that the contribution which is reflected at the interface $z = -d_2$ is 0.91% of the incident power for 150 MHz and 0.17% for 225 MHz. This value matches very well with those which can be read off from Fig.5.21. Consequently, as in hyperthermia, two contradictory requirements have to accounted for: low frequencies allow a large penetration depth, high frequencies allow the use of small antennas and — as will be discussed in section 6.5 — a good resolution.

The choice of the transmitting antenna is governed mainly by two aspects: what lateral resolution must be guaranteed, and how costly may the construction be? The use of dipole antennas requires little constructional expenditure and therefore is the cheapest solution. According to chapter 3 the input impedance can be determined approximately by supposing that the dipole is located within a homogeneous dielectric with a relative dielectric constant $(\varepsilon_{r1} + 1)/2$. The input impedance can then be calculated e.g. with the help of eqn.(6.24), given in the chapter 6. With this assumption the resonance length is about

$$\frac{\lambda}{2 \cdot Re\{\sqrt{(\varepsilon_{r1} + 1)/2}\,\}} \tag{5.29}$$

Often, for reasons of availability and cost–saving, dipole lengths below the resonance length will be used. Matching to the resonance length would mean that the antenna has to be replaced if ε_r changes. Concerning the accuracy of the model calculations this does not imply a complication, as the computer program takes the actual length of the antenna into account. From the experimental point of view the mismatch, which occurs more strongly out of resonance, certainly is disturbing.

Because of their low directivity, dipole antennas can only be used if lateral faults of the plane–layer–model are far away from the areas to be measured. A rough rule of thumb is that the undisturbed region ρ_s has to be approximately given by

$$\rho_s \gtrsim 2d_I \,, \tag{5.30}$$

where d_I denotes the depth of the interface to be measured.

The use of arrays of antennas substantially increases the numerical and experimental effort. For the calculations first of all Green's function has to. be supplied in the region $0 < \rho < \rho_{max}$, where ρ_{max} is the maximal extension of the array. As has been discussed, the number of integration points needed for the solution of the Sommerfeld integrals grows linear with ρ. The size of the systems of equations which have to be solved for the determination of the antenna currents grows linearly with the number of dipoles. This is exemplified by the computer time taken to prepare Figs. 5.3 and 5.9: 250 seconds and 2200 seconds respectively.

The determination of favourable values for $\Delta y, \Delta x$ (Fig.5.7) is simple only if the region above the interface to be measured is homogeneous (as in Figs. 5.8 and 5.9), or if it only shows little differences in the values for ε_{ri} so that one can use an effective ε_r as an approximation. In this case it turns out that — as for an array in free space — Δx and Δy should be slightly smaller than the wavelength in this stratum in order to obtain a small mainlobe (cf. Fig.5.8).

The optimal way in which model calculations should be set up depends substantially on the given task. With the help of the curves given in Figs. 5.11 to 5.17 it can be stated to what extent variations of the material parameters $\varepsilon'_{r2}, \varepsilon''_{r2}$ can be measured as variations of Z_{21}. This procedure makes sense e.g. if the depth or the size of a water–bearing layer has to be measured and all the other parameters of the stratified structure are known. Another possible application is the control of layer thicknesses and/or material parameters in industrial production processes. By careful determination of the relative positions of the transmitting and receiving antenna the sensitivity of the procedure concerning a certain parameter can be increased considerably. Thus one can read off e.g. from Fig.5.12 that the change of ε'_{r2} in the interval

$$2 \lesssim \varepsilon'_{r2} \lesssim 14, \quad \varepsilon''_{r2} = 1$$

produces a change of Z_{21} in the interval

$$-0.02\Omega \gtrsim R_{21} \gtrsim -0.10\Omega$$
$$0.12\Omega \gtrsim X_{21} \gtrsim -0.12\Omega$$

It can be seen from the table in section 5.3.3 (page 176) that $|Z_{21}|$ is smaller than $0.03\,\Omega$ for these positions of the transmitting and receiving dipoles and $d_1 \to \infty$, i.e. without the interface at $z = -d_1$. Thus $Z_{21}(d_1 \to \infty)$ is practically negligible. In the least favourable case 11 ($\Delta x = 0, \Delta y = 20$ cm), on the other hand, one has $|Z_{21}| = 0.76\,\Omega$. A change of ε'_{r2} in the range given above now produces a change of Z_{21} in the interval

$$-0.44\Omega \gtrsim R_{21} \gtrsim -0.56\Omega$$
$$-0.38\Omega \gtrsim X_{21} \gtrsim -0.61\Omega$$

The sensitivity is consequently substantially lower than in case 5.

A large number of methods for the solution of the inverse scattering problem, i.e. for the direct determination of material and geometry data from the measurement, are based on measurements with variable frequency. Model calculations as in Figs. 5.20 to 5.22 allow a reliable determination of the most favourable frequency range and the layout of the antennas. A further application of the model calculations in this context results from the fact that for the development of such solution procedures one needs data for the field distribution on the surface of the earth in order to test the validity of the method. Because of errors, experimental data are often useless, as one cannot decide whether deficiencies in the reconstruction are due to errors in measurement or to inadequacies of the reconstruction procedure. With the help of the procedure presented here, sets of data can be produced which are free of errors. By suitable modification of these sets of data the influence of single error sources can be investigated, whether they relate to the measurement apparatus or to data processing.

The same considerations also hold for procedures where, at constant frequency, the relative positions of the antennas are changed (tomography). With regard to the work required for the model calculations, this procedure is to be preferred, as now Green's function only has to be calculated once — even for very many values of ρ. For variable frequency, Green's function has to be recalculated for every frequency.

5.4 Summary of Chapter 5

An important aim of geological investigations with the help of electromagnetic waves is the determination of the distribution of material below the surface of the earth from measurements in the air above. As, up to now, it has not been possible to solve these inverse scattering problems satisfactorily, model calculations are needed for the interpretation of measured data. Further examples for the application of model calculations are the establishment of suitable measurement frequencies, the layout and the positioning of transmitting and receiving antennas, investigations for the assessment of the precision with which a certain quantity can be measured and the supply of sets of data in order to check the reconstruction procedures.

The materials of special interest in such investigations can be roughly classified into rocks, soil, ice and snow. The ε_r'-values, which can be found, are between about 1 and 14. ε_r'' has extremely small values for insulators and very high values for metals. In most applications, however, one is dealing with moderately damping dielectrics because of a high content of humidity.

The model calculations carried out in sections 5.3.1 and 5.3.2 address themselves to the question of the validity of the plane-layer-model. To this end the field within the structure to be investigated is calculated. For given measurement apparatus and precision, the region, which must be free of faults that cannot be described by the model, can be specified by means of the presentation of contours of constant level. As expected, it turns out that this region is very large if dipole antennas are used as transmitting antennas. Small regions are obtained when arrays are used, the design of which can be optimized by model calculations. Note that the calculation effort for arrays is very high.

In section 5.3.3 some model calculations for the coupling of two dipole antennas are discussed. Figs. 5.11 to 5.17 were produced under the assumption that the variation of the material parameters within an interface (here $z = -d_1$) and the depth of this interface are to be determined. The sensitivity of the procedure is enhanced by the fact that the transmitting and the receiving dipole are decoupled because of their relative positions, when the interface to be investigated is absent. It can be read off from the produced curves, how much the coupling is modified, if one of the parameters of interest changes. Thus the sensitivity of the procedure can be determined.

Calculations of the coupling as a function of frequency, as shown in Figs. 5.20 to 5.22, make it possible to present the influence of the single layers separately, if these are lossy and not too thin. Frequency dependent data in addition are suitable to test reconstruction procedures. As for every frequency point Green's function has to be recalculated, frequency dependent model calculations are very time consuming.

Tasks which come from the field of geology usually require the treatment of structures with many layers. An exhaustive discussion of the influence of all parameters therefore is not feasible. Even with only two, three or four layers - as in the examples discussed here - one is forced by the multitude of parameters to restrain oneself to a few exemplary investigations. Thus, systematic investigations of what information the crosspolar field components on the surface of the earth can give about the structure of the layers are still lacking. This statement corresponds to the fact that reconstruction procedures so far typically work with a scalar ansatz. Preliminary calculations of the crosspolar field components encourage the expectation that systematic investigations in this direction will be promising.

CHAPTER 6

Transient response of dipoles

6.1 The present state of research

The publications dealing with the analysis of the transient response of dipoles can roughly be divided into two groups: on the one hand we have determinations of the radiation properties of dipoles and wires of finite length in free space and on the other hand those of point sources (electric or magnetic Hertzian dipoles) in or above stratified media. As the method presented in this chapter is a combination of these two approaches, both groups will be discussed briefly.

The analysis dealing with transients on thin current filaments in free space which was mentioned first in the literature dates from 1923 [6.1]. Here the time-dependent current at each end of the wire is assumed to be the source of a spherical wave, and it is presumed that the current distribution as a function of the time is known. It was not until the sixties that further studies followed. Using approximate values for the input impedance and the effective length from [6.3], Schmitt [6.2] determined the step response of a thin dipole by means of numerical evaluation of the Fourier transform. Brundell [6.4] gives the exact solution (in the form of an infinite integral) for the field of a thin antenna of infinite length fed by a voltage impulse. Another solution, which Einarsson [6.5] has evaluated for some examples, also makes possible the determination of the time-dependent current distribution on the antenna. Obviously without knowing paper [6.4], the current distribution was also determined by Wu [6.6] and, based on this, by Morgan [6.7]. The evaluation of the solutions mentioned is very complicated without a high speed computer, because of the cylindrical functions in the integrand of an infinite integral. At that time high speed computers were not available. That is probably why the method introduced by Brundell in [6.4] was taken up again in 1970 in [6.8] and discussed in detail. Further results, based on Wu's solution, have been given in [6.9] by Kasevich. Simple approximations for the input impedance of Wu [6.10] were used by King and Schmitt [6.11] in order to discuss the early pulse response of a very long antenna. Bolle and Jacobs [6.12] determined the radiation field for the same antenna under the assumption that there exists only one travelling wave on the antenna. The determination of the transient fields of dipoles with finite length excited by a step function by Schmitt et al. in [6.13] is - similarly to what is done in [6.2] - based on approximate equations for the input impedance and the effective length derived in [6.3] and [6.14]. Since the development of high speed computers in the sixties there has been the opportunity to evaluate the Fourier transform also for complicated equations for the antenna currents or the radiating fields in the frequency domain. Thus, Harrison and King [6.15] determined the radiation field of an infinite cylindrical antenna excited by a Gaussian pulse using equations for the input impedance from King and Schmitt [6.11]. The work by Palciauskas and Beam [6.16] is based on equations for the radiation field of Wu [6.10] and on Hallen's solution for the input impedance of dipoles of finite length [6.17]. The transient response of two coupled dipole antennas is calculated in [6.18] and [6.19] starting with eqn. (3.135) with sinusoidal currents on both antennas and with input impedances equal to those of open-circuit loss-free transmission lines.

During the same time numerical methods were also derived for the solution of electromagnetic boundary problems in the frequency domain. Thus, it became possible in principle to analyse wire antennas with arbitrary shape. As this, however, involves much computer time, the first papers only deal with rather simple cases. For example, Landstorfer [6.20] determined the current distribution of a straight thin dipole with the method of moments in the frequency domain and then carried out the Fourier transform numerically. In the same manner [6.21] analyses how a band limited voltage source has to be chosen to maximize the field strength of the farfield at a specified farfield position and time. In [6.22] the method is extended to an array of dipoles. The time-dependent nearfield of a dipole excited by a rectangular pulse has been determined by Abo-Zena and Beam [6.23]. In [6.24] the properties of such dipoles are studied, for which the "Thin-Wire-Theory" is not valid any more. A good summary of the work published until the middle of the seventies is given by Franceschetti and Papas [6.25] and in the book "Transient electromagnetic fields" [6.26] edited by Felsen.

The determination of the current distribution in the frequency domain is much simpler for dipoles with nonreflecting resistive loadings. Then, only one wave which is excited by the feed is propagating, and the current distribution can be determined analytically [6.27]. Thus, particularly for this special case, the numerical determination of the transient fields becomes much easier ([6.28] and [6.29]).

It is a common feature of most of the papers mentioned here that first the frequency domain response of the structure is computed, and this is subsequently Fourier transformed to yield the time domain response. Towards the end of the sixties a method was developed which uses directly an integral equation for the electromagnetic field in the time domain [6.30]. Here the solution is obtained in the form of a time-stepping procedure, where the currents induced on a body at a time t are determined from the incident field and the current distribution at an earlier time $(t - \Delta t)$. Thus, the integral equation can be treated like an initial value problem. While Bennett and Weeks [6.30] analysed the diffraction of conductive bodies with large dimensions compared to the wavelength, Sayre and Harrington (for example [6.31]) dealt with the problem of wire antennas, which are of interest here; starting with the retarded vector potentials, they obtained two coupled integrodifferential equations. The derivation of further suitable integral equations and the discussion of numerical aspects are found in the papers [6.32] to [6.35]. The most important formulations used up to 1973 are reviewed by Poggio and Miller [6.36]. As examples of application, the work by Miller, Landt et al. ([6.37] and [6.38]) must be mentioned, where first the transient current distribution and then, from this via a Fourier transform, the input impedance in the frequency domain is determined. In [6.39] the nearfield, in [6.40] an array, is analysed in the time domain. The papers [6.41] and [6.42] present further summary work. An improvement of the 'marching-on in time procedure' is obtained by applying the method of conjugate gradient [6.43], [6.44]. In this method a suitably defined error is successively minimized at the end of each iteration. By this a solution is obtained after the error has become smaller than a certain preselected value.

At the beginning of the seventies the "Singularity Expansion Method", usually abridged as SEM, was developed. It was based on the finding that the transient response of various complicated antennas and scatterers appears to be dominated by a few damped sinusoids [6.45]. In order to determine these, the Laplace transforms of the time-dependent currents or of the fields are calculated and the singularities of these functions are determined ([6.46] and [6.47]). With the help of this method Pine and Tesche determined the early step response first for an infinite long wire [6.48], then for a wire of finite length [6.49]. In the years following quite a number of further papers about the SEM were published; here only those which deal with the radiation properties of straight dipoles or with basic research are quoted ([6.50] to [6.57]). In the book "Transient electromagnetic fields" edited by Felsen, Baum [6.58] published an excellent introduction to the concept which the SEM is based on. Further information can be found in a special edition of the periodical "Electromagnetics" of 1981 [6.59].

The determinations of the transient response of sources in plane stratified media were at first limited to the simplest cases, those of electric or magnetic Hertzian dipoles in an interface between two half spaces of different dielectrics. One of the first papers on this topic, that of van der Pol [6.60] as early as 1956, presents the exact solution for the important special case of an electric dipole, placed vertically on the interface and excited by a step function. The observation point also lies in the interface. Because both half spaces are assumed to be lossfree, both integrations - that for the representation of the field in the frequency domain and that for the inverse Laplace transform - can be performed on the real axis. Another paper, which was published almost at the same time [6.61], is based on the plane waves representation of the field. The results, however, can only be interpreted for some very special setups of the source point and the observation point. A paper based on the numerical evaluation of one of the integrals was published in 1957 by Pekeris and Alterman [6.62]. Here, after some analytical manipulations, it was possible to obtain integrals with finite intervals of integration. In an extension to the paper of van der Pol [6.60] by Bremmer [6.63] the field is described by contributions which travel along different paths - above, below and on the earth's surface - from the transmitter to the receiver. Johler and Walters carried out the Fourier transform for the solution of Norton [2.36] with damped sinusoidal source currents [6.64]. The papers of Wait ([6.65] and [6.66]) are based on Sommerfeld's approximate solution. He discusses the field of a dipole which is due to an excitation by a ramp function. The inverse transform is performed numerically as well as with the help of approximate equations valid for the case of high conductivity of the earth. From the beginning of the sixties onwards a great number of papers follow, which also deal with the problem of a point source above an earth which is taken as homogeneous (for example [6.67] to [6.73]). A summary of the papers published up to the end of the seventies, dealing with the application of transient electromagnetic waves in geological prospecting is given by Lee [6.74]. A solution proposed for the case of two different layers can be found in a paper by Botros and Mahmoud [6.75]. For the determination of the early pulse response, they start with a solution in the frequency domain, which is in accordance with the saddle-point solution (eqns. (2.100) to (2.105)) and thus with an interpretation in terms of geometrical optics. For a source above a homogeneous lossy earth it is shown by Haddad and Chang [6.76] that the numerical integration can be made considerably simpler by deforming the path of integration, analogous to the procedure of the SEM. A similar procedure is given by Ezzedine et al. [6.77] for two layers without losses and later on [6.78] for lossy layers.

All the papers which have been quoted so far deal with the case which is of interest here, namely that of an electric Hertzian dipole. At the same time, work on the properties of magnetic dipoles has also been published, and the development of these methods was similar to that concerning electric dipoles. This is why these papers will only be mentioned in summary ([6.79] to [6.87]).

A lot of work was published in the context of the "Induced Polarization Method". This method for measuring the electrical properties of the surface of the earth implies the use of time dependent excitations, where the upper limit of the frequency is typically several hertz. The corresponding mathematical methods, therefore, cannot be applied to the problems that will be treated in this chapter.

Concerning the related problem of a pulsed line source of infinite length above a stratified media only some recently published papers will be mentioned. In [6.88] to [6.90] the transient field of a line source above a plane conducting half space is obtained as a finite integral over the transient plane wave solution for complex angles of incidence. A parametric study was performed in [6.91] and

[6.92].

Up to now only a few papers have been published which deal with the transient response of finite dipole antennas above multilayered structures, with dipole lengths not very small compared to the wavelength at the upper limit of the frequency range. With the help of the SEM, Rech et al. [6.93] have analysed the radiation properties of log-periodic antennas, consisting of dipoles normal to the earth. In [6.94] Rahmat-Samii et al. and in [6.95] Taylor and Crone have treated the analogous problem with wire antennas arranged horizontally above the earth. In both cases the earth was assumed to be a homogeneous half space.

In the present work we shall discuss the case of a dipole of finite length, which is excited by a signal of limited bandwidth and which is located above a plane structure consisting of any number of layers. To this end we make use of the results in the frequency domain of chapter 3. The solution in the time domain is obtained by Fourier transformation. In order to limit the numerical effort, in the next section only observation points in the farfield will be considered. We shall introduce suitable approximations.

6.2 Comments on the approximations used

The calculations are based on the equivalent circuit of Fig. 6.1. The generator is specified by the internal resistance R_g and the open-circuit voltage $U_g(f)$. Z_{in} is the frequency dependent input impedance of a dipole. Then - based on equation (2.17) - the equation for the radiation field in the observation point $\vec{r}$ in the frequency domain reads

$$\vec{E}(\vec{r},f) = U_g(f)\frac{1}{R_g + Z_{in}(f)}\int\limits_{2l}\vec{G}(\vec{r},\vec{r}',f)\cdot\vec{u}_x\frac{I_x(x',f)}{I_x(0,f)}dx' \qquad . \qquad (6.1)$$

Formally the solution in the time domain can be obtained via Fourier transform (small letters are used for the time domain):

$$\vec{e}(\vec{r},t) = \int\limits_{-\infty}^{+\infty}\vec{E}(\vec{r},f)e^{j2\pi ft}df \qquad . \qquad (6.2)$$

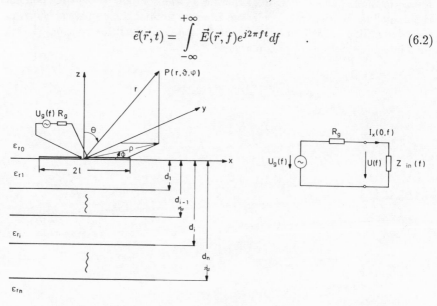

Fig.6.1   A dipole above stratified earth and the equivalent circuit

Although all terms required in eqn. (6.1) can be determined with the methods described in chapters 2, A2 and A3, a numerical analysis is extremely expensive and because of this is often not practicable, because for each frequency one has to

1) determine the Green's function for the range $0 \leq \rho \leq \rho_{max}$, where ρ_{max} denotes the maximum overall size of the antenna,

2) use the method of moments to determine the current distribution and the input impedance,

3) determine the Green's function for the observation point,
4) perform the integration over the current distribution.

Therefore, the purpose of the following considerations is to search for approximations, which lead to manageable equations for the cases of technical interest.

Measurements of transient responses, for example in the context of geological explorations, must always be carried out within a limited frequency range. In order to receive a strong response from a band-limited pulse, in the frequency domain the pulse must have a high amplitude close to the first resonance frequency of the antenna used. This requirement can be fulfilled for example by a modulated Gaussian pulse, described by

$$U(f) = U_{go}\sqrt{\frac{1}{2\pi}}\frac{\sigma}{2}\left(e^{-2(\pi\sigma(f+f_m))^2} + e^{-2(\pi\sigma(f-f_m))^2}\right) \qquad (6.3)$$

or

$$u(t) = u_{go}\cos(\omega_m t)\,e^{-t^2/(2\sigma^2)} \qquad (6.4)$$

with the centre frequency f_m being equal to the first resonance frequency of the antenna.

The parameter σ will be chosen in such a way that in the frequency domain the amplitude of $U_g(f)$ for an upper limit f_{max} is equal to 5% of the amplitude for the centre frequency f_m. Thus, for σ one obtains the equation

$$\sigma = \sqrt{1.5}\,\frac{1}{\pi(f_{max} - f_m)}\Big|_{U_g(f_{max})=0.05U_g(f_m)} \qquad . \qquad (6.5)$$

When using those or similar band-limited signals the system function need only be known in the range $|f| < f_{max}$ for sufficient accuracy.

For the determination of $I_x(x', f)$ and $Z_{in}(f)$ the results of section 3.3.1 will be used. There it was found that the relative current distribution on a dipole in a stratified medium is approximately equal to that on a dipole in a homogeneous space with an effective ε_{re}, where ε_{re} is equal to the arithmetic mean of ε_{r0} and ε_{r1} (eqn. (3.99)). In the next section a suitable method for the determination of the current distribution in homogeneous space is given (eqn. (3.97) used in section 3.3.2 obviously cannot be applied here). It follows from eqn. (6.1) that incorrect values for the input impedance $Z_{in}(f)$ can falsify the result to a large extent. Therefore $Z_{in}(f)$ will be determined not only according to the same approximate method used for the current distribution but also with the help of the method of moments.

As a further simplification only observation points in the farfield of the antenna above the layers will be considered, so that eqns. (2.100) to (2.105), which were obtained in section 2.3.2 via the saddle-point method, can be used. Denoting the required components of the Green's tensor by

$$G_{\vartheta,\varphi} = \frac{dE_{\vartheta,\varphi s}^{TM/TE}}{I_x dl_x} = G_{\vartheta,\varphi}^{\infty}(\vartheta,\varphi,f)\frac{e^{-jkr}}{r} \qquad , \qquad (6.6)$$

equation (6.1) can be written as

$$E_{\vartheta,\varphi}(\vec{r},f) = U_g(f)\frac{1}{R_g + Z_{\text{in}}(f)}G_{\vartheta,\varphi}^{\infty}(\vartheta,\varphi,f)\frac{e^{-jkr}}{r}\int\limits_{2l}\frac{I_x(x',f)}{I_x(0,f)}e^{+jk(r-r')}dx'.$$

(6.7)

Thus under the three stated presuppositions the system function consists of three parts: the excitation, the Green's function - which depends only on the parameters of the layers, the coordinates of the observation point and the frequency - and the relative current distribution which depends on the dipole length and the dielectric constants of the immediate surroundings of the dipole:

$$E_{\vartheta,\varphi}(r,\vartheta,\varphi,f) = U_g(f)\cdot H_1(f)\cdot H_2(f)$$

(6.8)

with

$$H_1(f) = \frac{1}{R_g + Z_{\text{in}}(f)}\int\limits_{2l}\frac{I_x(x',f)}{I_x(0,f)}e^{+jx'k\sin\vartheta\,\cos\varphi}\,dx' \qquad ,$$

(6.9)

$$H_2(f) = G_{\vartheta,\varphi}^{\infty}(\vartheta,\varphi,f)\frac{e^{-j\frac{2\pi f}{c}r}}{r} \qquad .$$

(6.10)

6.3 The frequency domain solution

6.3.1 Approximate equations for the current distribution and the input impedance

A suitable method for the determination of the current distribution on a dipole in homogeneous space is the modified King-Middleton solution given in [2.9]. The range in which it is valid is defined by

$$2l < \frac{3}{2} \frac{\lambda_{min}}{Re\{\sqrt{\varepsilon_{re}}\}} \qquad (6.11)$$

The equation for the current is (according to [2.9] p. 161)

$$I(x) = \frac{j2U\pi}{Z_e \Psi_{kR}\cos(k_e l)}[\sin(k_e(l-|x|)) + T_k(\cos(k_e x) - \cos(k_e l))] \qquad . \ (6.12)$$

Here Ψ_{kR} and T_k are functions of the frequency f and the length of the dipole $2l$, but not of the source point x'. Z_e is equal to the intrinsic impedance of a space with the relative dielectric constant ε_{re}:

$$Z_e = \sqrt{\frac{\mu_0}{\varepsilon_0 \varepsilon_{re}}} \ . \qquad (6.13)$$

Ψ_{kR} is determined by a numerical solution of the integral

$$\Psi_{kR}(u) = \frac{1}{\sin(k_e(l-|u|))} \qquad (6.14)$$

$$\int\limits_{-l}^{+l} \sin(k_e(l-|x'|)) \left[\frac{\cos(\beta_e r_u)\cosh(\alpha_e r_u)}{r_u} - \frac{\cos(\beta_e r_l)\cosh(\alpha_e r_l)}{r_l} \right] dx'$$

with

$$k_e = \frac{2\pi f}{c}\sqrt{\varepsilon_{re}} \ , \qquad (6.15)$$

$$\beta_e = Re\{k_e\} \qquad \text{and} \qquad \alpha_e = -Im\{k_e\}, \qquad (6.16)$$

$$r_u = \sqrt{(u-x')^2 + a^2} \ , \qquad r_l = \sqrt{(l-x')^2 + a^2} \ . \qquad (6.17)$$

The ranges $\beta_e l \leq \pi/2$ and $\beta_e l > \pi/2$ must be treated in different way corresponding to the prescription

$$\Psi_{kR}(u) = \begin{cases} \Psi_{kR}(0) & \text{for } \beta_e l \leq \frac{\pi}{2} \\ \Psi_{kR}(l - \frac{\lambda_e}{4}) & \text{for } \beta_e l > \frac{\pi}{2} \end{cases} \qquad . \qquad (6.18)$$

The correction term T_k is obtained from

$$T_k = T_k(l) = -\frac{(1 - \cos(k_e l))\Psi_{Vk} - j\Psi_{kI}\cos(k_e l)}{(1 - \cos(k_e l))\Psi_{Uk} - \Psi_{kU}\cos(k_e l)}. \tag{6.19}$$

The terms Ψ_{Vk}, Ψ_{Uk}, Ψ_{kI} and Ψ_{kU} of this equation must be calculated using the following integrals

$$\Psi_{Vk} = \int_{-l}^{+l} \sin(k_e(l - |x'|)) \frac{e^{-jk_e r_l}}{r_l} dx' \quad, \tag{6.20}$$

$$\Psi_{Uk} = \int_{-l}^{+l} (\cos(k_e x') - \cos(k_e l)) \frac{e^{-jk_e r_l}}{r_l} dx' \quad, \tag{6.21}$$

$$\Psi_{kI} = \Psi_{kI}(u = 0)$$
$$= \int_{-l}^{+l} \sin(k_e(l - |x'|)) \tag{6.22}$$

$$\left[\frac{\sin(k_e r_l) - \sin(\beta_e r_l)\sinh(\alpha_e r_l)}{r_l} - \frac{\sin(k_e r_u) - \sin(\beta_e r_u)\sinh(\alpha_e r_u)}{r_u} \right] dx'$$

$$\Psi_{kU} = \Psi_{kU}(u = 0) = \int_{-l}^{+l} (\cos(k_e x') - \cos(k_e l)) \left[\frac{e^{-jk_e r_u}}{r_u} - \frac{e^{-jk_e r_l}}{r_l} \right] dx' \tag{6.23}$$

By eqn. (6.12) we get for the input impedance

$$Z_{\text{in}}(f) = -j\frac{Z_e \Psi_{kR}\cos(k_e l)}{2\pi[\sin(k_e l) + T_k(f)(1 - \cos(k_e l))]} \quad. \tag{6.24}$$

With eqn. (6.12) the integral in eqn. (6.7) can be solved analytically. With the definition of the function $F(\vartheta, \varphi)$ from eqn. (3.96) one obtains

$$F(\vartheta, \varphi) = \int_{-l}^{+l} I_x(x')e^{jkx'\sin\vartheta \cos\varphi} dx' \tag{6.25}$$

$$= \frac{j2\pi U_e}{Z_e \Psi_{kR}\cos(k_e l)} \{ (\frac{1}{k_+} - \frac{1}{k_-})(\cos(k\eta l) - \cos(k_e l))$$
$$+ T_k[\frac{1}{k_-}\sin(k_- l) + \frac{1}{k_+}\sin(k_+ l) - \frac{2}{k\eta}\cos(k_e l)\sin(k\eta l)]\},$$

where the following abbreviations have been used:

$$\eta = \sin\vartheta \cos\varphi \quad, \tag{6.26}$$

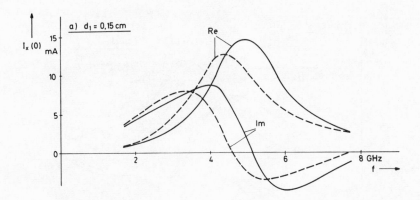

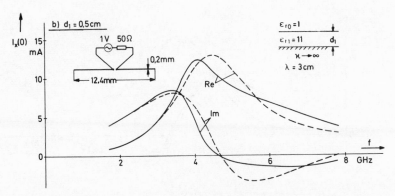

Fig.6.2 Input current $I_x(0)$ of a dipole on a grounded dielectric sheet ——————
and in homogeneous media with $(\varepsilon_{r1}+1)/2$ ------

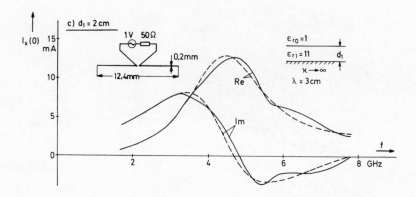

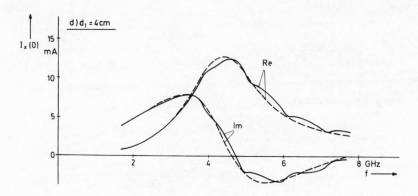

Fig.6.2 Input current $I_x(0)$ of a dipole on a grounded dielectric sheet ————
and in homogeneous media with $(\varepsilon_{r1}+1)/2$ ------

$$k_+ = k\eta + k_e \quad , \tag{6.27}$$

$$k_- = k\eta - k_e \quad . \tag{6.28}$$

In order to find out how well the current distributions and the input impedances can be described by eqns. (6.12) and (6.24), numerous examples have been calculated, both using these equations and numerically. These show that eqn. (6.12) - in agreement with the results displayed in the Figs. 3.18 and 3.19 - allows a determination of the integral in eqn. (6.7) to an adequate accuracy in all cases investigated. The precise error can only be given for each concrete case. For the term

$$I_x(0, f) = \frac{U_g(f)}{R_g + Z_{\mathrm{in}}(f)} \tag{6.29}$$

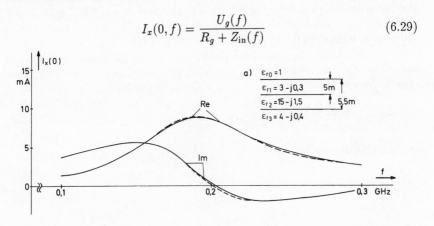

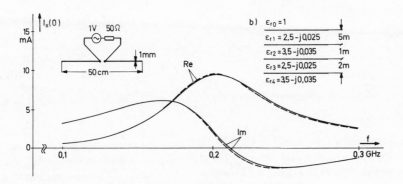

Fig.6.3 Input current $I_x(0)$ of a dipole on stratified media ——— and in homogeneous media with $(\varepsilon_{r1}+1)/2$ ‒‒‒‒‒

however, considerable deviation can occur between the results of the two procedures. As an example Figs. 6.2a-d show the results for a lossless dielectric layer with $\varepsilon_{r1} = 11$ and different thicknesses of the layer on a perfect reflector. The dipole length is 1.24 cm, the frequency range is 1.8 to 7.8 GHz. For a small thickness of 0.15 cm (Fig. 6.2a) the approximate eqn. (6.24) leads to useless results. This is also true for all structures which at present are typically used for microstrip antennas (for example ceramic substrate with $\varepsilon_{r1} = 10$ and $d_1 = 0.0635$ cm). In the case of thicker layers the error becomes substantially smaller (Figs. 6.2b to 6.2d). The curves determined numerically oscillate round those obtained using the approximate equation, and the frequency of this oscillation increases as the thickness of the layer grows. Already small losses within the layers lead to a substantial decrease in the deviations. Figs. 6.3a and 6.3b show two examples from the field of geology (section 5.3.3) . In both cases the differences related to the maximum value are less then 2%.

In the literature concerning the application of the method of moments, it is noted that the calculation of the currents can become incorrect for high frequencies (for example [6.96]). By the restriction to band-limited signals with

$$f_{max} \leq \frac{c}{2l\,\mathrm{Re}\{\sqrt{\varepsilon_{re}}\,\}} \tag{6.30}$$

the calculations are carried out in a range which leads to reliable results.

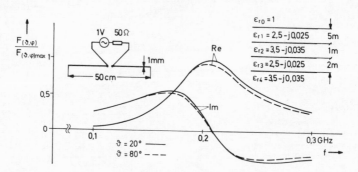

Fig.6.4 Function $F(\vartheta,\varphi = 45°)$ (eqn. 6.25) for $\vartheta = 20°$ and $\vartheta = 80°$

It has already been noted in section 3.3.2 that the factor $\exp(jkx'\sin\vartheta\,\cos\varphi)$ in (6.25) can be set equal to unity for short dipoles. This in indeed the case here, as the dipoles are operated close to the first resonant frequency. Fig. 6.4, as an example, shows the function $F(\vartheta,\varphi)$ for the angles $\vartheta = 20°$ and $\vartheta = 80°$ with a constant $\varphi = 45°$. The parameters of the layers are equal to those of Fig. 6.3b. One can see that both curves show some deviations, especially in the upper frequency range.

6.3.2 The saddle-point solution

Eqns. (2.100) to (2.105) immediately yield $G_{\vartheta,\varphi}$:

$$G_{\vartheta}(f) = -j\cos\varphi \,\frac{f\mu_0}{2}\left[1 - \Gamma_0^{TM}(f,\vartheta)e^{-j4\pi f\frac{d_0}{c}\cos\vartheta}\right]\frac{e^{-j2\pi fr/c}}{r}\cos\vartheta \qquad (6.31)$$

and

$$G_{\varphi}(f) = j\sin\varphi \,\frac{f\mu_0}{2}\left[1 + \Gamma_0^{TE}(f,\vartheta)e^{-j4\pi f\frac{d_0}{c}\cos\vartheta}\right]\frac{e^{-j2\pi fr/c}}{r} \qquad (6.32)$$

where c is the velocity of light in free space.

For further discussions of the saddle-point solution the abbreviation introduced in eqn. (6.6)

$$G_{\vartheta,\varphi}^{\infty}(\vartheta,\varphi) = \frac{G_{\vartheta,\varphi}}{e^{-jkr}/r} \qquad (6.33)$$

is used. In free space $G_{\vartheta,\varphi}^{\infty}$ is overall proportional to the frequency; in the case of layered media in addition to this the reflections at the various interfaces have to be taken into account. This becomes apparent from Fig. 6.5, where $G_{\vartheta,\varphi}^{\infty}$ is shown for $\varphi = 45°$ and some angles ϑ. The parameters for Fig. 6.5 are the same as those for Fig. 6.3b. Because of the fact that the dielectric constants ε_{r1} to ε_{r4} differ from each other only slightly reflections are very weak. Therefore, the contributions of the different interfaces are not discernible, as already shown in chapter 5.

One arrives at a presentation which allows for a more intuitive interpretation using the Bremmer series [6.97]. Thus, due to this series, the waves described by the reflection coefficients $\Gamma_0^{TM/TE}$ can be split up into parts which are reflected several times at the interfaces (Fig. 6.6) [6.98]. Here, just as in [6.99], for simplicity a two-layer structure will be analysed in some detail, where the dipole lies on the first layer, i.e. $d_0 = 0$.

By the expansion of the reflection coefficients $\Gamma_0^{TM/TE}$ in a power series and a simple manipulation of the eqns. (A2.12) and (A2.13) one obtains

$$\Gamma_0^{TM/TE}(\vartheta,f) = r_0^{TM/TE} \qquad (6.34)$$

$$+ \left(r_0^{TM/TE} - \frac{1}{r_0^{TM/TE}}\right)\sum_{q=1}^{\infty}(-1)^q \left(r_0^{TM/TE}r_1^{TM/TE}\right)^q \exp\left(-j2qk_{z1}d_1\right)$$

where r_0 and r_1 are given by

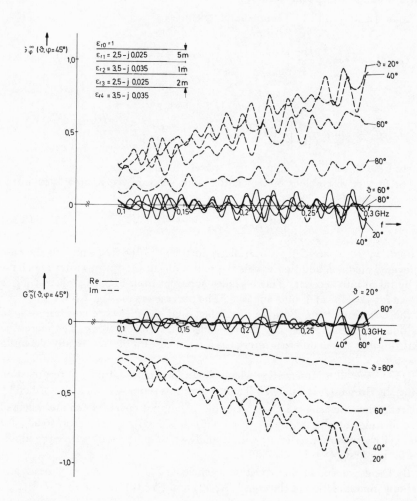

Fig. 6.5 Saddle-point solution $G_{\vartheta,\varphi}^{\infty}(\vartheta,\varphi=45°)$ in the frequency domain

$$r_i^{TM} = \frac{\varepsilon_{r(i+1)}k_{zi} - \varepsilon_{ri}k_{z(i+1)}}{\varepsilon_{r(i+1)}k_{zi} + \varepsilon_{ri}k_{z(i+1)}} \quad , \tag{6.35}$$

$$r_i^{TE} = \frac{k_{zi} - k_{z(i+1)}}{k_{zi} + k_{z(i+1)}} \quad , \tag{6.36}$$

$$k_{zi} = k\sqrt{\varepsilon_{ri} - \sin^2\vartheta} \quad , \quad i = 0,1 \quad . \tag{6.37}$$

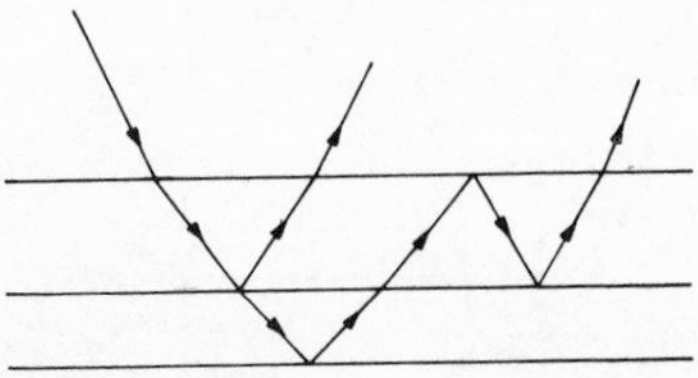

Fig. 6.6 Multiple reflections at the interfaces

For a microstrip antenna without a cover one obtains with $|\varepsilon''_{r2}| \to \infty$

$$\Gamma_0^{TM/TE}(\vartheta, f) = \frac{r_0^{TM/TE} + / - e^{-2jk_{z1}d_1}}{1 + / - r_0^{TM/TE}e^{-2jk_{z1}d_1}} \quad , \tag{6.38}$$

or, as a series representation according to eqn. (6.34)

$$\Gamma_0^{TM/TE}(\vartheta, f) = r_0^{TM/TE} + \left(r_0^{TM/TE} - \frac{1}{r_0^{TM/TE}} \right)$$
$$\sum_{q=1}^{\infty}(-/+1)^q(r_0^{TM/TE})^q \exp\left(-j2\pi f\frac{2qd_1}{c}\sqrt{\varepsilon_{r1} - \sin^2\vartheta} \right) \quad . \tag{6.39}$$

If now eqn. (6.37) is inserted into eqns. (6.31) and (6.32) we get

$$G_\vartheta^\infty(\vartheta, \varphi) = -j\cos\varphi\, \frac{f\mu_0}{2}\cos\vartheta\, (1 - e^{-2jk_{z1}d_1})(1 - r_0^{TM})$$
$$\sum_{q=0}^{\infty}(-1)^q(r_0^{TM})^q e^{-2jqk_{z1}d_1} \tag{6.40}$$

and

$$G_\varphi^\infty(\vartheta, \varphi) = j\sin\varphi\, \frac{f\mu_0}{2}(1 - e^{-2jk_{z1}d_1})(1 + r_0^{TE})$$
$$\sum_{q=0}^{\infty}(r_0^{TE})^q e^{-2jqk_{z1}d_1} \quad . \tag{6.41}$$

As an example, in Fig. 6.7 the curves for $G_{\vartheta,\varphi}^\infty$ are given for the structure of Fig. 6.2c ($\varepsilon_{r1} = 11, d_1 = 2$ cm). They show that the results obtained now are very different from those in Fig. 6.5. Because of the factor

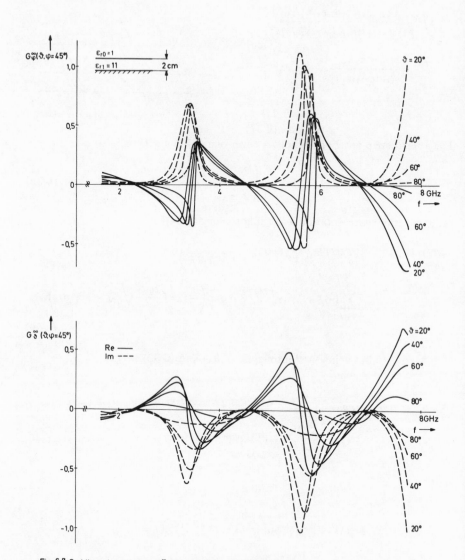

Fig. 6.7 Saddle-point solution $G_{\vartheta\varphi}^{\infty}(\vartheta, \varphi = 45°)$ in the frequency domain

$$(1 - e^{-2jk_{z1}d_1}) \tag{6.42}$$

for the dielectric media without losses zeros of the real part of $G^{\infty}_{\vartheta,\varphi}$ are obtained for

$$f^{TM/TE}_{NR} = \frac{N_R c}{2d_1 \sqrt{\varepsilon_{r1} - \sin^2\vartheta}} \quad ; \qquad N_R = 0, 1, 2, \ldots \tag{6.43}$$

and of the imaginary part for

$$f^{TM/TE}_{NI} = \frac{N_I c}{4d_1 \sqrt{\varepsilon_{r1} - \sin^2\vartheta}} \quad ; \qquad N_I = 0, 1, 2, \ldots \quad . \tag{6.44}$$

In the case of dielectric media with losses the zeros of the imaginary part change to

$$f^{TM/TE}_{NI} = \frac{N_I c}{4d_1 \cdot Re\{\sqrt{\varepsilon_{r1} - \sin^2\vartheta}\}} \quad . \tag{6.45}$$

The zeros of the real part become minima. In the case of $\varepsilon_{r1} \neq f(\omega)$ the factors

$$(1 - r_0^{TM}) \quad \text{and} \quad (1 + r_0^{TE})$$

are independent of the frequency. The convergence of the series in eqns.(6.41) and (6.42) is determined by the reflection factors at the interface, $r_0^{TM/TE}$, and in the case of lossy material also by $Im\{k_{z1}\}$ and by the thickness of the layer d_1. The result shown in Fig. 6.7 can be easily transferred to other thicknesses of the layer by compressing or expanding the frequency axis. For $d_2 = 0.15$ cm (compare Fig. 6.2a) the first zero of the imaginary part lies at approximately 30 GHz, which is well outside the range $1,8 - 7,8$ GHz shown here. In the case of $d_2 = 4.0$ cm, however, we have five zeros within this range.

6.3.3 The system function

The system function is obtained in the frequency domain from the product stated in eqn. (6.8)

$$U_g(f) \cdot H_1(f) \cdot H_2(f) \quad . \tag{6.46}$$

Here — following the restrictions stated in section 6.2 — the centre frequency of the excitation $U_g(f)$ is chosen approximately equal to the first resonance frequency of the dipoles. The upper limit is determined by the following considerations: On the one hand the range in which the King–Middleton solution is valid is limited by the non-equality (6.11). More important in this context, however, is the fact that a dipole, which for the first resonance frequency is matched to a source, is totally mismatched round about the second resonance frequency. This is why the upper limit is chosen in such a way that the range round about the second resonance frequency is not taken into account for the calculations; the fixing is not very critical. The lower limit then results from

$$f_{min} = 2f_m - f_{max} \quad .$$

(6.47)

The numerical experiments discussed in the following were performed for the following data:

a) dielectric layer with reflector:

$$f_m = 4.75 \text{ GHz}$$
$$f_{max} = 7.80 \text{ GHz}$$
$$\Delta f = 2(f_{max} - f_m) = 6.10 \text{ GHz}$$

(6.48)

b) geological stratified structures:

$$f_m = 0.20 \text{ GHz}$$
$$f_{max} = 0.30 \text{ GHz}$$
$$\Delta f = 2(f_{max} - f_m) = 0.20 \text{ GHz}$$

(6.49)

Beyond the band limits $U_g(f)$ is set equal to zero. The calculated examples show that this leads to a negligible error.

The result of the multiplication of $U_g(f)$ and $H_1(f)$ can easily be overlooked: The function is mainly determined by $H_1(f)$ with the range around the first resonance frequency being increased and that towards the band limits being decreased. From this it follows that $H_1(f)$, and with this the input impedance, must be very carefully determined. The procedure chosen here guarantees this because the King–Middleton method as well as the numerical solution works particularly reliably in this case.

Performing the multiplication with $H_2(f)$ (eqn. (6.10)) several cases can be studied, which will be discussed following the examples shown in section 6.3.2. In the case of a thin, lossfree grounded dielectric sheet ($d_1 < 0.5$ cm) $H_2(f)$ possesses only one zero in the frequency range specified by the source function. Thus the result of the product eqn. (6.8) becomes very involved and can no longer be generally discussed. In the case of thick layers, however, $H_2(f)$, and thus also $E_{\vartheta,\varphi}(\varphi, \vartheta, f)$, possess several zeros in the range $f_{min} < f < f_{max}$ that we are interested in. Therefore the system function is mainly determined by the rapidly varying function $H_2(f)$. In the case of geological structures it results from the multiplication of the curves shown in Figs. 6.3 and in 6.5 that now the frequency dependence of a dipole in homogeneous space dominates and that the layers are only noticeable as small oscillations.

6.4 Transient response of dipoles

6.4.1 The dipole in homogeneous space

From the inverse Fourier transform of the product $U_g(f) \cdot H_1(f)$

$$h_{g1}(t) = \mathcal{F}^{-1}\{U_g(f) \cdot H_1(f)\} \tag{6.50}$$

we obtain the impulse response of a dipole in homogeneous space, if the input impedance $Z_{in}(f)$ is determined with eqn. (6.24). A simple qualitative discussion is only possible in special cases, e.g. when the pulse length is much smaller than the time delay

$$\tau_l = \frac{l \cdot \sqrt{\varepsilon_{re}}}{c} \tag{6.51}$$

which a pulse needs for the distance between the feed area and the dipole ends (e.g. [6.25]). In this case the pulse response can approximately be divided into a part of the feed area and two parts of the dipole ends, which arrive delayed by

$$\tau_l(1 \pm \cos\varphi \, \sin\vartheta \,) \quad . \tag{6.52}$$

Simple approximations can also be derived for the case of very short antennas ($l \ll c/f_{max}$ with f_{max} equal to the frequency at the upper band limit). One can see that the radiated field is proportional to the second derivation of $U_g(f)$ with respect to time [6.26].

Neither of the two special cases is valid here, because, if one specifies the pulse width in the time domain by the -20 dB–values, and applies the band limits given in eqns. (6.46) and (6.47), for the pulse length of the excitation $\Delta\tau_{pe}$

$$\Delta\tau_{pe} = \frac{3.35}{\Delta f/\,\mathrm{GHz}}\,\mathrm{nsec} \tag{6.53}$$

is obtained. In case a (Fig. 6.2) (a dielectric layer with reflector) we get

$$\Delta\tau_{pe} = 0.55\,\mathrm{nsec} \quad , \tag{6.54}$$

in case b (Fig. 6.3) (geological stratified structures)

$$\Delta\tau_{pe} = 16.75\,\mathrm{nsec} \quad . \tag{6.55}$$

For the time delay τ_l with the dipole length of Fig. 6.2 (case a)

$$\tau_l = 0.05\,\mathrm{nsec} \tag{6.56}$$

and with those of Fig. 6.3 (case b)

$$\tau_l = 1.18\,\mathrm{nsec} \tag{6.57}$$

are obtained. This is why the pulse response of a dipole in homogeneous space
has also to be calculated numerically with the help of a FFT–algorithm. For
the modulated Gaussian pulse used here as excitation only a slight change of
the shape of the pulse is obtained with a small expansion of the pulse length
(denoted by $\Delta\tau_p$).

6.4.2 The saddle-point solution

By carrying out the inverse Fourier transform of the saddle-point solutions
(6.31) and (6.32) the impulse responses of the stratified structure can be calculated. One obtains (with the dielectric constants of all layers being real and
independent of the frequeny)

$$g_\vartheta(t) = -\cos\varphi\, \frac{\mu_0}{4\pi}\frac{1}{r}\delta(t - \frac{r}{c}) * \frac{\partial}{\partial t}[\delta(t) - \gamma_0^{TM}(t - 2\frac{d_0}{c}\cos\vartheta\,)]\cos\vartheta \quad (6.58)$$

$$g_\varphi(t) = \sin\varphi\, \frac{\mu_0}{4\pi}\frac{1}{r}\delta(t - \frac{r}{c}) * \frac{\partial}{\partial t}[\delta(t) + \gamma_0^{TE}(t - 2\frac{d_0}{c}\cos\vartheta\,)] \quad (6.59)$$

where $\gamma_0^{TM/TE}$ are the inverse Fourier transforms of the reflection coefficients.

A particularly simple equation for the impulse response of a microstrip structure without losses results from the series representation (6.38). Using eqn.
(6.38) first of all the inverse Fourier transforms of the reflection factors are obtained

$$\gamma_0^{TM/TE}(t) = r_0^{TM/TE}\delta(t) + (r_0^{TM/TE} - \frac{1}{r_0^{TM/TE}})$$

$$\sum_{q=1}^{\infty}(-/+1)^q (r_0^{TM/TE})^q \delta(t - \frac{2qd_1}{c}\sqrt{\varepsilon_{r1} - \sin^2\vartheta}\,) \quad . (6.60)$$

Insertion of (6.43) into (6.41) and (6.42) results in

$$g_\vartheta(t) = -\cos\varphi\, \frac{\mu_0}{4\pi r}\delta(t - \frac{r}{c}) * \frac{\partial}{\partial t}\{\delta(t) - r_0^{TM}\delta(t) - (r_0^{TM} - \frac{1}{r_0^{TM}})$$

$$\sum_{q=1}^{\infty}(-1)^q (r_0^{TM})^q \delta(t - \frac{2qd_1}{c}\sqrt{\varepsilon_{r1} - \sin^2\vartheta}\,)\}\cos\vartheta \quad , \quad (6.61)$$

$$g_\varphi(t) = \sin\varphi\, \frac{\mu_0}{4\pi r}\delta(t - \frac{r}{c}) * \frac{\partial}{\partial t}\{\delta(t) + r_0^{TE}\delta(t) + (r_0^{TE} - \frac{1}{r_0^{TE}})$$

$$\sum_{q=1}^{\infty}(-1)^q (r_0^{TE})^q \delta(t - \frac{2qd_1}{c}\sqrt{\varepsilon_{r1} - \sin^2\vartheta}\,)\} \quad . \quad (6.62)$$

These equations are readily interpreted: an impulse exciting a microstrip structure without losses is radiated with a delay of multiples of the time

$$\tau_d = \frac{2d_1}{c}\sqrt{\varepsilon_{r1} - \sin^2\vartheta} \quad . \quad (6.63)$$

The amplitudes which are governed by the reflections at the metal reflector and the interface between air and dielectric can be calculated from the factor

$$A_q^{TM/TE} = (r_0^{TM/TE} - \frac{1}{r_0^{TM/TE}})(-/ + r_0^{TM/TE})^q \qquad . \qquad (6.64)$$

The considerations made so far have been based on the assumption that the materials are dispersion–free and thus e.g. $r_0^{TM/TE}$ is independent of the frequency. According to the Kramers–Kronig relation this is not strictly possible; however, in the case of lossfree dielectric media it is certainly a very good approximation. In the case of highly lossy materials, however, the calculation of the pulse response, assuming that there is no dispersion, typically leads to results which do not satisfy the principle of causality. If one puts

$$\varepsilon_{ri}(\omega) = \varepsilon'_{ri}(\omega) - j\varepsilon''_{ri}(\omega) \qquad (6.65)$$

then $r_0^{TM/TE}$ and k_{zi} (eqns. (6.35) to (6.37)) become complex and become functions of frequency. Thus in eqns. (6.61) and (6.62) instead of multiplying with $r_0^{TM/TE}$ convolution integrals have to be carried out. The inverse transform of the term

$$exp\left(-2j\pi f\frac{2qd_1}{c}\sqrt{\varepsilon_{r1}(\omega) - sin^2\vartheta}\right) = exp(e_q) \qquad (6.66)$$

can only be formed numerically in the case of frequency dependency, as is, for example, required in the context of geophysical problems (chapter 5). On the other hand a simple analytical solution can be achieved when small losses can be described by a frequency independent conductivity:

$$\varepsilon_{r1}(\omega) = \varepsilon'_r - j\frac{\kappa_1}{\omega\varepsilon_0} \qquad \text{with} \qquad \frac{\kappa}{\omega\varepsilon_0} \ll 1 \qquad . \qquad (6.67)$$

Thus we gain

$$\mathcal{F}^{-1}(exp(eq)) = \delta\left(t - \frac{2qd_1}{c}\sqrt{\varepsilon'_{r1} - sin^2\vartheta}\right) exp\left(-\frac{1}{2}\frac{2qd_1}{c\varepsilon_0}\frac{\kappa}{\sqrt{\varepsilon'_{r1} - sin^2\vartheta}}\right) \qquad . \qquad (6.68)$$

This expression can be easily interpreted: because of the losses the pulses are dampened, but their shape is not changed.

6.4.3 The transient response of dipoles on stratified media

The time domain solution is obtained from the frequency domain via inverse Fourier transform (eqn. (6.2)). Using eqns. (6.50), (6.58) and (6.59) the solution can also be written as a convolution integral

$$e_{\vartheta,\varphi}(t) = \int\limits_{-\infty}^{+\infty} h_{g1}(\tau) \cdot g_{\vartheta,\varphi}(t-\tau)d\tau = h_{g1}(t) * g_{\vartheta,\varphi}(t) \quad . \tag{6.69}$$

To avoid formal difficulties with the interpretation of the convolution integral, the derivation with respect to time occurring in eqns. (6.61) and (6.62) is applied to the term $h_{g1}(\tau)$. Generally $e_{\vartheta,\varphi}(t)$ has to be calculated with the help of a FFT–algorithm. Using the Bremmer series one can, however, give a formulation which allows a qualitative interpretation of the result. Here again we limit our discussions to lossfree microstrip structures. One obtains

$$e_{\vartheta}(t) = -\cos\varphi\,\cos\vartheta\,\frac{\mu_0}{4\pi}\cdot\frac{1}{r}\frac{\partial}{\partial t}$$
$$(h_{g1}(t)) * \left[(1 - r_0^{TM})\delta(t - \frac{r}{c}) - (r_0^{TM} - \frac{1}{r_0^{TM}}) \right. \tag{6.70}$$
$$\left. \sum_{q=1}^{\infty}(-1)^q r_0^{TM}\delta(t - \frac{r}{c} - q\cdot\tau)\right]$$

$$e_{\varphi}(t) = \sin\varphi\,\frac{\mu_0}{4\pi}\cdot\frac{1}{r}\frac{\partial}{\partial t}$$
$$(h_{g1}(t)) * \left[(1 + r_0^{TE})\delta(t - \frac{r}{c}) + (r_0^{TE} - \frac{1}{r_0^{TE}}) \right. \tag{6.71}$$
$$\left. \sum_{q=1}^{\infty}(-1)^q r_0^{TE}\delta(t - \frac{r}{c} - q\cdot\tau)\right]$$

Thus we get a series of pulses in the form

$$\frac{\partial}{\partial t}(h_{g1}(t)) \quad . \tag{6.72}$$

The amplitude of the pulses rapidly decreases with a small magnitude of the reflection factor at the interface dielectric - air and/or in the case of high attenuation inside the material. The pulse can only distinctively be seen separately when the interval between the pulses, τ_d, is larger than the pulse length, $\Delta\tau_p$. Here we have to remember that the bandwidths of the excitation $u_g(t)$ and $U_g(\omega)$ respectively were defined in section 6.3.3 by

$$0.5 f_{res1} \lesssim f \lesssim 1,5 f_{res1} \tag{6.73}$$

and thus by the lengths of the dipoles. Therefore short pulses can only be achieved with short dipole antennas and correspondingly high frequencies. The interval between the pulses τ_d is fixed – following eqn. (6.63) – by the layer thickness and ε_{r1}. This way one can easily determine the required minimum dipole length for a given thickness of the layer.

In the case of transient problems it is common to talk about an early time and a late time response. The early time response is mostly influenced by the high-frequency content of the pulse, or more precisely speaking, by frequency parts, for which the corresponding wave lengths are much smaller than the geometrical dimensions. For the late time response just the opposite is valid. For the case

$$\tau_d \gg \Delta\tau_p \qquad\qquad (6.74)$$

the pulse response can be called an early time response with respect to the stratified structure. Therefore its solution can also be obtained with the help of geometrical optics. As far as the dipole is concerned here we have neither an early nor a late time response, because the wave length corresponding to the centre frequency is of the same order as the dipole length.

The dependence of the pulse response on φ (xy-plane) is given by $cos\varphi$ or $sin\varphi$ respectively. The ϑ- dependence is on the one hand determined by $h_{g1}(t)$ — i.e. by the length of the dipole — especially, however, by the dependence on ϑ of the reflection factors. In addition to this, the ϑ–component has to be multiplied with $cos\vartheta$. A quantitative statement is not possible because of the convolution integrals.

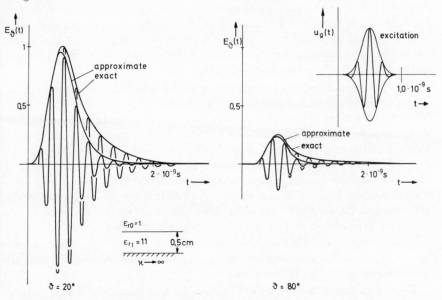

Fig. 6. 8 Comparison of the exact and the approximate (based in equ. 6.24) solution for the impulse response ($\varphi=45°$)

6.4.4 Discussion of some examples

In this section the results for the structures discussed so far (see Figs. 6.2 and 6.3) are presented. The Gaussian pulse is specified by the parameters given in section 6.3.3.

First of all those errors will be considered which occur due to using eqn. (6.24) for the calculation of the input impedance. As an example Fig. 6.8 shows the results obtained by using the approximate eqn. (6.24) and those obtained via the method of moments for a dielectric layer of 0.5 cm thickness with a reflector (see Fig. 6.2 b). Note that eqn. (6.24) leads to a too narrow pulse response. As was to be expected, for the structure of Fig. 6.2 a ($\varepsilon_{r1} = 11, d_1 = 0.15$ cm) eqn. (6.24) leads to useless results. In the case of the dielectric layers of 2 cm and 4 cm thickness (for example Fig. 6.2 c and 6.2 d), however, the difference between the results of the two different procedures is smaller than the thickness of the line; the same holds for the two geological structures of Figs. 6.3 a and 6.3 b.

The impulse responses for all structures discussed so far are shown in Figs. 6.9 to 6.14 for the angles $\vartheta = 20°, 40°, 60°$ and $80°$. There it is supposed that all dielectric constants can be taken as independent of frequency in the frequency range under consideration. Figs. 6.11 and 6.12 are easiest to interpret: they show a series of pulses at an interval of approximately $0.43 \cdot 10^{-9}$s and $0.86 \cdot 10^{-9}$s, respectively; this value is almost independent of ϑ in agreement with eqn. (6.63). The first pulse results from the superposition of the pulse radiated directly from the source and the one reflected at the interface $z = 0$. Its amplitude is proportional to $|1 - r_0^{TM}|$ or $|1 + r_0^{TE}|$, respectively. For certain angles this amplitude can be smaller than the amplitude of the second pulse, which is equal to $|(r_0^{TM/TE})^2 - 1|$ according to eqn. (6.64).

Because the material was taken as being free of dispersion the envelope of the responses and those of the excitations differ only slightly. For $d_1 = 0.5$ cm (Fig. 6.10) the time delay of the pulses is now only $0.11 \cdot 10^{-9}$ s and thus much shorter than the pulse length $\Delta \tau_p$. It follows from this that the pulse response can no longer be divided into separate pulses. The same also applies particularly to Fig. 6.9 with $d_1 = 0.15$ cm. Further we see in this figure that the envelope of the pulse response has a minimum which comes about $0.25 \cdot 10^{-9}$ s later than the maximum. This minimum cannot – unlike before – be explained by the reflections at the ground or at the interface substrate/air but is caused by the frequency dependence of the input impedance (Fig. 6.2 a).

It follows from the results just discussed that the microstrip structures with $\varepsilon_{r1} = 10, d_1 = 0.635$ mm nowadays frequently used in the microwave range are hardly suitable for a pulse operation.

Figs. 6.13 and 6.14 show pulse responses for geological stratified structures (Figs. 6.3 a and b). In Fig. 6.14 the reflections at the lower layers can easily be identified, whereby the contributions of the single interfaces are superposed

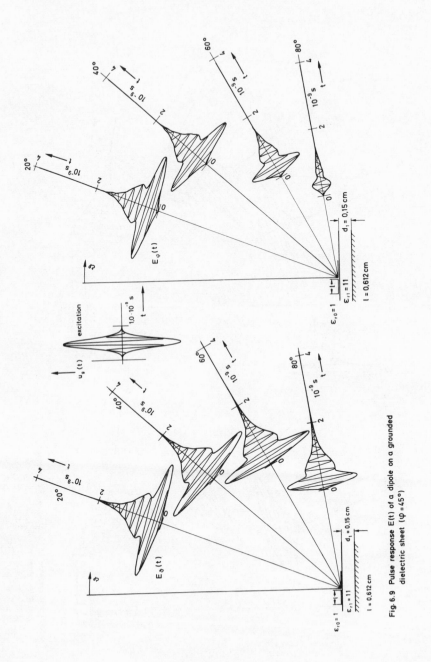

Fig. 6.9 Pulse response E(t) of a dipole on a grounded
dielectric sheet (φ = 45°)

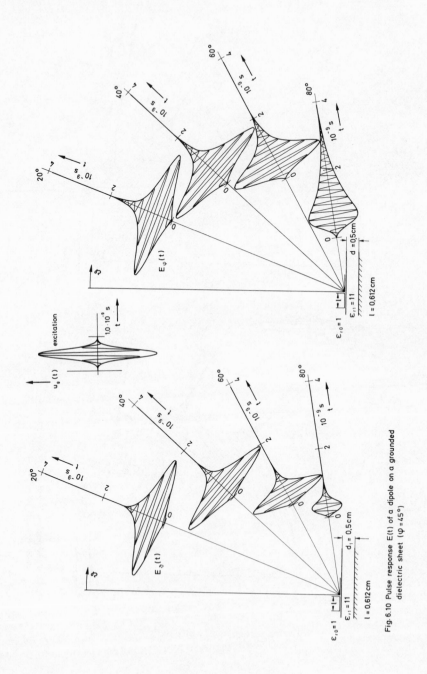

Fig. 6.10 Pulse response E(t) of a dipole on a grounded dielectric sheet ($\varphi = 45°$)

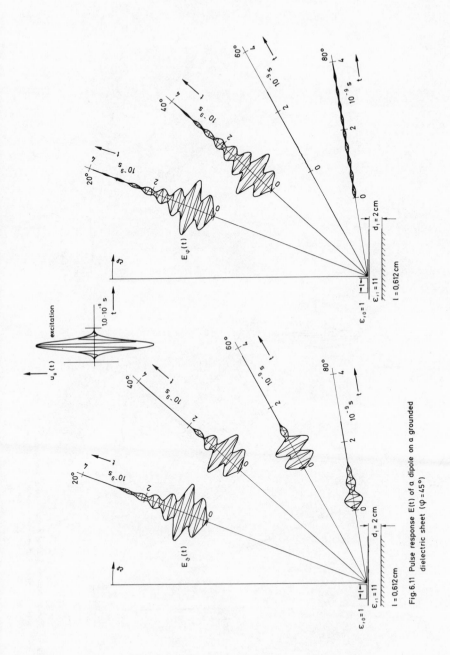

Fig. 6.11 Pulse response E(t) of a dipole on a grounded dielectric sheet (φ = 45°)

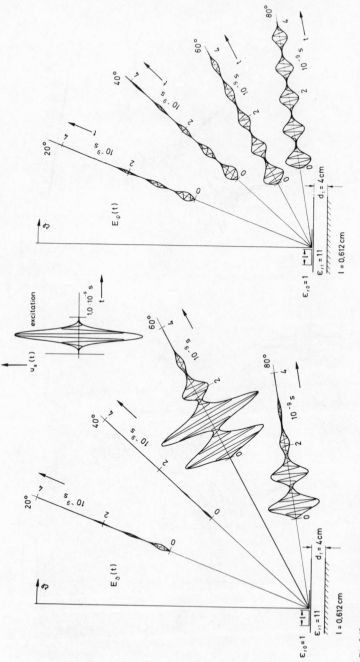

Fig. 6.12 Pulse response E(t) of a dipole on a grounded dielectric sheet ($\varphi = 45°$)

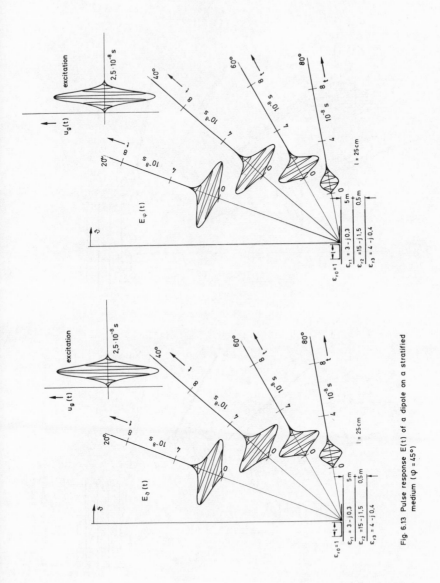

Fig. 6.13 Pulse response E(t) of a dipole on a stratified medium ($\varphi = 45°$)

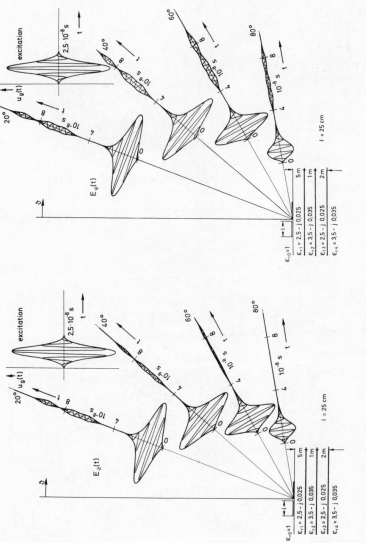

Fig. 6.14 Pulse response E(t) of a dipole on a stratified medium (φ = 45°)

due to the small thickness of the second layer. The layered structure of Fig. 6.13 differs from the others in its high dielectric losses. Even the amplitude of the pulse which is reflected at the interface $z = -5$ m and reradiated is strongly damped. Therefore, it is not discernible in the linear scale used here.

6.5 Summary of Chapter 6

The calculation of a pulse response is extremely complicated, because first the integral equation for the antenna current in the frequency range has to be solved, and then an inverse Fourier transform has to applied. To achieve a practical method despite these facts, the following restrictions must be observed:
- the excitation is a modulated Gaussian pulse, the centre frequency being equal to the first resonance frequency of the dipole f_{res1}, and the bandwidth approximately equal to $1,5 f_{res1}$,
- the normalized current distribution of the dipole located on the stratified structure is set equal to that of a dipole in a homogeneous space with $(\varepsilon_{ro} + \varepsilon_{r1})/2$,
- The observation point lies in the farfield of the dipole.

With these restrictions the pulse response in the frequency range can be stated as a product of the excitation, the normalized field strength of a dipole in homogeneous space and of the Green's function of the stratified structure. For the calculation of the normalized current distribution the modified King–Middleton method is used. With this method the input impedance can also be determined, if the first layer is sufficiently thick. In the case of a thin first layer the input impedance has to be determined by solving the integral equation. The integrals occurring with the calculation of the field in homogeneous space can be performed analytically.

For the Green's function of a stratified structure the saddle-point solution is used. It consists of the contribution of the source itself and the contribution reflected at the surface. The latter can be developed in a Bremmer series. The Green's function of a lossfree dielectric microstrip structure without dielectric cover is discussed in detail.

An approximate determination of the pulse response of a dipole in homogeneous space can only be analytically performed easily if the pulse length is much shorter or much longer than the time delay, which the pulse needs for the distance between the feed area and the ends of the dipole. Because neither of these two cases is given here, the solution has to be obtained numerically. From this it results that the shape of the response differs only slightly from that of the excitation; only the pulse length increases slightly.

The time domain response of the saddle-point solution can very easily be stated with the help of the Bremmer series for the case of a lossfree microstrip structure. One obtains a series of impulses, the intervals of which are defined by the time delay required for the distance between air-substrate and ground. A simple calculation of the impulse response for lossy dielectric materials is only possible for a very special case. Otherwise a numerical calculation of the inverse Fourier transform has to be carried out.

At the end of chapter 6 some numerically calculated examples for the response of dipoles fed by modulated Gaussian pulses on microstrip structures and on geological structures are shown. Only if the pulse length is smaller than

the time delay of the pulses, can the pulses be easily identified and thus analysed. Formally one can always specify dipole length with which the short pulses
required can be excited. In the case of lossy layers, however, this method can
lead to two conflicting requirements: As was shown several times in the chapters 4 and 5 large penetration depths can only be achieved with low frequencies,
while high frequencies as required for the short pulses are highly dampened in a
lossy medium. Thus there is a restriction on the application of electromagnetic
methods.

APPENDIX

A1 Application of the method of moments for the solution of the integral equation

Consider the integral equation (2.21) in the abbreviated form

$$L[I_s] = F \quad , \tag{A1.1}$$

where L is a linear operator. As a first step I_s is expanded in a series of properly chosen basis functions f_p ([A1.1]):

$$I_s(\vec{r}) = \sum_{p=1}^{P} \hat{I}_p f_p(\vec{r}) \quad . \tag{A1.2}$$

In this equation the $\hat{I}_p$ are constants yet unknown. Because of the fact that L is a linear operator, the substitution of eqn. (A1.2) in (A1.1) results in

$$\sum_{p=1}^{P} \hat{I}_p L[f_p(\vec{r})] = F \quad . \tag{A1.3}$$

Then a set of Q weighting functions $w_1, w_2, ..., w_Q$ is chosen in the domain of L and (A1.3) is multiplied with each w_i ($i = 1, ..., Q$) after an appropriate inner product has been defined. If, as usual, the inner product of two functions ϕ and ψ is denoted by $< \phi, \psi >$ one obtains the set of equations

$$\sum_{p=1}^{P} \hat{I}_p < w_q, L[f_p] > = < w_q, F > , q = 1, ..., Q \quad . \tag{A1.4}$$

According to Harrington ([A1.2]), in the case we are dealing with here, the integral

$$< \phi, \psi > = \int_C \phi(x)\psi(x)dx \ , x \in C \tag{A1.5}$$

is a suitable inner product of the functions $\phi(x)$ and $\psi(x)$; C denotes the range for which the solution is required. Thus, the integration has to be performed over the complete printed circuit. If one sets $P = Q$ then with (A1.4) we have a set of P equations for the determination of the P unknown $\hat{I}_p$. The special choice

$$f_p = w_q \quad \text{for} \quad p = q \tag{A1.6}$$

is known as Galerkin's method.

It is also possible, as Richmond [A1.3] showed, to derive the set of equations (A1.4) from the reaction concept formulated by Rumsey [A1.4].

The question which basis and weighting functions are suitable has been discussed in detail for a wire antenna in homogeneous space (see e.g. [A1.5] to [A1.8]). We distinguish between functions which are well-defined and non-zero in the entire domain of the operator L (entire domain functions) and functions which are defined in the domain of L but are zero in parts of the domain (subdomain functions). In the first published numerical determination of the current distribution on a straight thin wire antenna with the method of moments the antenna current was expanded in cosine functions as entire domain basis functions [A1.9].

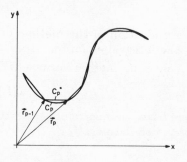

Fig. A1.1 Approximation of a curved segment c by a straight segment c

As, for example, is shown in [A1.2] and [A1.10], in general the subdomain basis functions which will be used here are more flexible in their application. For their definition the printed circuit, described by C, is divided into P segments (Fig. A1.1). The segments are called $C_p, p = 1, ..., P$ where the p'th segment is determined by the vectors $\vec{r}_{p-1}$ and $\vec{r}_p$. Curved parts of C are approximated by straight segments; this new contour is called C^*. For the analysis of wire antennas in free space sinusoidal subdomain bases are often used, defined by the equations

$$f_p(\vec{r}) = U_p \frac{\sin k|\vec{r} - \vec{r}_{p-1}|}{\sin k|\vec{r}_p - \vec{r}_{p-1}|} + U_{p+1} \frac{\sin k|\vec{r}_{p+1} - \vec{r}|}{\sin k|\vec{r}_{p+1} - \vec{r}_p|} \tag{A1.7}$$

with

$$U_p = \begin{cases} 1 & ,\vec{r} \in C_p^* \\ 0 & ,\vec{r} \notin C_p^* \end{cases} \tag{A1.8}$$

This approach has the advantage that $L[f_p]$ can be calculated analytically.

For the selection of the weighting functions it is generally very important to take into account that the definition of the inner product requires a further integration over C or C^* respectively. In order to avoid tedious calculations subdomain functions are almost exclusively used. If one uses piecewise sinusoidal weighting functions (Galerkin's method), then for antennas in free space the coefficients of the set of equations (A1.4) can be determined in a simpler way with some analytical considerations. Apart from this, the set of equations can then be interpreted in physical terms. To this end, two adjacent segments C_p and C_{p+1} are considered to be one dipole D_p. The term

$$\hat{I}_p L[f_p] \qquad (A1.9)$$

represents the electric field of the dipole D_p tangential to C, in each point on C. If, further, one takes two adjacent segments to be the dipole D_q with the current distribution $\hat{I}_q w_q$, then the term

$$<w_q, L[f_p]> = \int_{D_q} w_q \left(\vec{s}(\vec{r}) \int_{D_p} \vec{G}(\vec{r}, \vec{r}') \cdot s(\vec{r}') f_p(\vec{r}') ds' \right) ds = -Z_{pq} \quad (A1.10)$$

is equal to the negative mutual coupling between the dipoles D_p and D_q [A1.11]. The right-hand side of the q th equation of (A1.4) is not equal to zero only if the q th dipole contains a voltage source. If this source lies between two segments and if S^+/w can be assumed to be small (Fig. 2.3) the right-hand side represents the voltage source.

For the stratified structures which are analysed in this book, generally the coefficients of eqn. (A1.4) can only be calculated numerically. The determination of the Z_{pp} provides special difficulties, because the integrands for distances approaching zero between source- and field-points become singular. In section 2.4.4 approximate equations have been derived with the help of which the field close to the source in a stratified structure is described by image sources in a homogeneous space. Then, using piecewise sinusoidal basis and weighting functions - as given in eqn. (A1.7) - simplified equations for the determination of the self-impedances which have been developed for antennas in homogeneous space can be used here as well.

A2 The reflection coefficients

In each of the $(\hat{m} + n + 1)$ layers the equations for the field are set up with
the still unknown amplitudes A_i, B_i, C_i and D_i in the form given by eqns.(2.62)
to (2.73). These amplitudes have to be determined in such a way that both the
TM- and the TE-part of the field fulfil the boundary conditions. It is assumed
that a current element is oriented parallel to the x-axis and that it is situated
in the origin of the coordinate system:

$$\vec{I}dl = \vec{u}_x I_x dl_x \quad . \tag{A2.1}$$

In view of the discussion of some
applications we shall specialize the
structure of Fig. 2.1 according to
Fig. A2.1. We assume $| \varepsilon_{r\hat{1}} | \to \infty$
for layer $\hat{1}$, which means that the
dipole is screened towards the upper half space [2.104]. The procedure used in the following is closely
related to the one discussed by
Kong ([2.78]) in the context of the
calculation of Green's function of
an unscreened stratified structure.
We shall proceed in such a way
that an extension to the original
structure of Fig. 2.1 is straightforward.

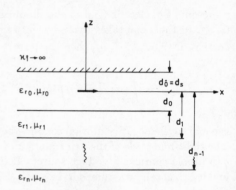

Fig. A2.1 Stratified medium with reflector at $z = d_s$

The wave amplitudes in the interface at $z = -d_i$, namely $\tilde{a}_i$ and $\tilde{b}_i$, or $\tilde{c}_i$ and
$\tilde{d}_i$ respectively, are obtained from the wave amplitudes in the layer i , namely
A_i and B_i or C_i and D_i respectively, as follows:

$$\tilde{a}_i(k_\rho) = e^{-jk_{zi}d_i}A_i(k_\rho) \quad , \tag{A2.2}$$

$$\tilde{c}_i(k_\rho) = e^{-jk_{zi}d_i}C_i(k_\rho) \quad , \tag{A2.3}$$

$$\tilde{b}_i(k_\rho) = e^{+jk_{zi}d_i}B_i(k_\rho) \quad , \tag{A2.4}$$

$$\tilde{d}_i(k_\rho) = e^{+jk_{zi}d_i}D_i(k_\rho) \quad . \tag{A2.5}$$

The ratio between the amplitudes of the reflected and the incident wave is determined by the boundary conditions. For TM-waves we get

$$\frac{\tilde{b}_i}{\tilde{a}_i} = \frac{\varepsilon(+)_i^{i+1}}{\varepsilon(-)_i^{i+1}} \left[1 - \frac{\dfrac{\varepsilon(+)_i^{i+1}}{\varepsilon(-)_i^{i+1}} - \dfrac{\varepsilon(-)_i^{i+1}}{\varepsilon(+)_i^{i+1}}}{\dfrac{\varepsilon(+)_i^{i+1}}{\varepsilon(-)_i^{i+1}} + \dfrac{\tilde{b}_{i+1}}{\tilde{a}_{i+1}}e^{-j2k_{z(i+1)}(d_{i+1} - d_i)}} \right] \tag{A2.6}$$

and for TE-waves

$$\frac{\tilde{d}_i}{\tilde{c}_i} = \frac{\mu(+)_i^{i+1}}{\mu(-)_i^{i+1}} \left[1 - \frac{\dfrac{\mu(+)_i^{i+1}}{\mu(-)_i^{i+1}} - \dfrac{\mu(-)_i^{i+1}}{\mu(+)_i^{i+1}}}{\dfrac{\mu(+)_i^{i+1}}{\mu(-)_i^{i+1}} + \dfrac{\tilde{d}_{i+1}}{\tilde{c}_{i+1}} e^{-j2k_{z(i+1)}(d_{i+1} - d_i)}} \right] \qquad (A2.7)$$

with

$$i = 0, 1, \ldots n \quad , \qquad\qquad\qquad (A2.8)$$

where the following abbreviations have been used:

$$\varepsilon(\pm)_q^p = \frac{\varepsilon_{rp}}{\varepsilon_{rq}} \pm \frac{k_{zp}}{k_{zq}} \quad , \qquad\qquad (A2.9)$$

$$\mu(\pm)_q^p = \frac{\mu_{rp}}{\mu_{rq}} \pm \frac{k_{zp}}{k_{zq}} \quad , \qquad\qquad (A2.10)$$

$$k_{zi} = \sqrt{k^2 \varepsilon_{ri} \mu_{ri} - k_\rho^2} , \qquad k^2 = \omega^2 \varepsilon_0 \mu_0 \ . \qquad (A2.11)$$

The thus defined reflection coefficients of the interface $z = -d_i$ will subsequently be referred to as $\Gamma_i^{TM}(k_\rho)$ or $\Gamma_i^{TE}(k_\rho)$, respectively, so that the equations

$$\Gamma_i^{TM}(k_\rho) = \frac{\tilde{b}_i}{\tilde{a}_i} = \frac{\varepsilon(-)_i^{i+1} + \varepsilon(+)_i^{i+1} \Gamma_{i+1}^{TM}(k_\rho) e^{-j2k_{z(i+1)}(d_{i+1} - d_i)}}{\varepsilon(+)_i^{i+1} + \varepsilon(-)_i^{i+1} \Gamma_{i+1}^{TM}(k_\rho) e^{-j2k_{z(i+1)}(d_{i+1} - d_i)}}, \quad (A2.12)$$

$$\Gamma_i^{TE}(k_\rho) = \frac{\tilde{d}_i}{\tilde{c}_i} = \frac{\mu(-)_i^{i+1} + \mu(+)_i^{i+1} \Gamma_{i+1}^{TE}(k_\rho) e^{-j2k_{z(i+1)}(d_{i+1} - d_i)}}{\mu(+)_i^{i+1} + \mu(-)_i^{i+1} \Gamma_{i+1}^{TE}(k_\rho) e^{-j2k_{z(i+1)}(d_{i+1} - d_i)}} \quad (A2.13)$$

follow after rewriting eqns. (A2.6) and (A2.7).

Since the area n is a source-free half space unlimited in the $(-z)$ direction we have

$$\tilde{b}_n = \tilde{d}_n = 0 \qquad\qquad\qquad (A2.14)$$

and thus

$$\Gamma_n^{TM} = \Gamma_n^{TE} = 0 \ . \qquad\qquad\qquad (A2.15)$$

In addition, the solution of the boundary value problem has to satisfy the boundary condition

$$\vec{E}_t(z = d_s) = 0 \qquad\qquad\qquad (A2.16)$$

in the region $z > 0$. This boundary condition can also be expressed with the help of reflection coefficients $\Gamma_s^{TM/TE}$ which connect the amplitudes of the waves reflected by the interface at $z = d_s$ with the amplitudes of the waves that are incident on the interface from the region "0".

For the general plane stratified medium according to Fig. 2.1 the $\Gamma_s^{TM/TE}$ follow from eqns. (A2.12) and (A2.13) with $d_i \to d_{\hat{i}}$ and $d_{\hat{0}} = d_s$. If the space $z > d_s$ is an unlimited half space of perfectly conducting material (Fig. A2.1), then we have $\hat{m} = 1$ and because of $\kappa_{\hat{i}} \to \infty$ formally $|\varepsilon_{r\hat{i}}| \to \infty$. Then eqns. (A2.9) to (A2.13) yield

$$\Gamma_s^{TM}(k_\rho) = +1 \qquad\qquad (A2.17)$$

and

$$\Gamma_s^{TE}(k_\rho) = -1 \quad . \qquad\qquad (A2.18)$$

A3 Equations for the wave amplitudes

A3.1 Equations in the region $-d_0 \leq z \leq +d_s$

In the range below the source $(-d_0 \leq z \leq 0)$ the waves running in the $(-z)$ direction consist of a part from the source itself - which will be denoted by A_{0-} and C_{0-} - and a part reflected by the upper screen - which will be denoted by A_r and C_r:

$$A_0(k_\rho) = A_{0-}(k_\rho) + A_r(k_\rho) \quad , \tag{A3.1}$$

$$C_0(k_\rho) = C_{0-}(k_\rho) + C_r(k_\rho) \quad . \tag{A3.2}$$

For all waves propagating from the interface d_0 towards the $(+z)$ direction holds:

$$B_0(k_\rho) = B_r(k_\rho) \quad , \tag{A3.3}$$

$$D_0(k_\rho) = D_r(k_\rho) \quad . \tag{A3.4}$$

However, the ratios of the amplitudes B_0/A_0 or D_0/C_0, respectively, are equal to the reflection coefficients Γ_0^{TM} or Γ_0^{TE} (valid at $z = -d_0$) if these have been transformed into the plane $z = 0$. Therefore the following relations hold:

$$B_r(k_\rho) = e^{-j2k_{z0}d_0} \, \Gamma_0^{TM}(k_\rho) \, [A_{0-}(k_\rho) + A_r(k_\rho)] \quad , \tag{A3.5}$$

$$D_r(k_\rho) = e^{-j2k_{z0}d_0} \, \Gamma_0^{TE}(k_\rho) \, [C_{0-}(k_\rho) + C_r(k_\rho)] \quad . \tag{A3.6}$$

(For this argument it was assumed that $Re\{k_{z0}\} > 0$, cf. section 2.4.1).

Based on similar considerations the equations

$$A_0(k_\rho) = A_r(k_\rho) \quad , \tag{A3.7}$$

$$C_0(k_\rho) = C_r(k_\rho) \quad , \tag{A3.8}$$

$$B_0(k_\rho) = A_{0+}(k_\rho) + B_r(k_\rho) \quad , \tag{A3.9}$$

$$D_0(k_\rho) = C_{0+}(k_\rho) + D_r(k_\rho) \tag{A3.10}$$

are valid for the region $0 \leq z \leq d_s$. The ratios of the amplitudes A_0/B_0 or C_0/D_0, respectively, result in this region from the reflection factors $\Gamma_s^{TM/TE}$ (valid at $z = d_s$) if these have been transformed into the plane $z = 0$. Thus

$$A_r(k_\rho) = e^{-j2k_{z0}d_s} \, \Gamma_s^{TM} \, [A_{0+}(k_\rho) + B_r(k_\rho)] \quad , \tag{A3.11}$$

$$C_r(k_\rho) = e^{-j2k_{z0}d_s} \, \Gamma_s^{TE} \, [C_{0+}(k_\rho) + D_r(k_\rho)] \quad . \tag{A3.12}$$

By means of these equations the wave amplitudes A_0 to D_0 can be expressed as functions of the amplitudes of the source, $A_{0\pm}$ and $C_{0\pm}$. One obtains for the region $-d_0 \leq z \leq 0$:

$$A_0 = \frac{1 + \frac{A_{0\pm}}{A_{0-}} R_s^{TM}}{1 - R_0^{TM} R_s^{TM}} A_{0-} \quad , \tag{A3.13}$$

$$C_0 = \frac{1 + \frac{C_{0\pm}}{C_{0-}} R_s^{TE}}{1 - R_0^{TE} R_s^{TE}} C_{0-} \quad , \tag{A3.14}$$

$$B_0 = R_0^{TM} A_0 \quad , \tag{A3.15}$$

$$D_0 = R_0^{TE} C_0 \quad . \tag{A3.16}$$

For the region $0 \leq z \leq d_s$ one obtains

$$B_0 = \frac{1 + \frac{A_{0-}}{A_{0+}} R_0^{TM}}{1 - R_0^{TM} R_s^{TM}} A_{0+} \quad , \tag{A3.17}$$

$$D_0 = \frac{1 + \frac{C_{0-}}{C_{0+}} R_0^{TE}}{1 - R_0^{TE} R_s^{TE}} C_{0+} \quad , \tag{A3.18}$$

$$A_0 = R_s^{TM} B_0 \quad , \tag{A3.19}$$

$$C_0 = R_s^{TE} D_0 \quad . \tag{A3.20}$$

In these equations $R_0^{TM/TE}$ and $R_s^{TM/TE}$ represent the reflection coefficients $\Gamma_0^{TM/TE}$ and $\Gamma_s^{TM/TE}$ transformed into the $z = 0$ plane, thus

$$R_{0,s}^{TM/TE} = e^{-j2k_{z0}d_{0,s}} \, \Gamma_{0,s}^{TM,TE} \quad . \tag{A3.21}$$

The easiest way to determine $A_{0\pm}$ and $C_{0\pm}$ is to write the E_z- and H_z-components of the current element $I_x dl_x$ in the homogeneous space with the help of the identity (2.38). By comparison with the approach according to eqns. (2.62) to (2.73) one obtains

$$A_{0\pm}(k_\rho) = \pm \frac{I_x dl_x k_\rho^2}{j8\pi\omega\varepsilon_0\varepsilon_{r0}} \quad , \tag{A3.22}$$

$$C_{0\pm}(k_\rho) = \frac{I_x dl_x k_\rho^2}{j8\pi k_{z0}} \quad . \tag{A3.23}$$

Since the stratified medium was assumed to be homogeneous with respect to ρ and φ the dependence of the field on φ is equal to that of a point source in a homogeneous space. Therefore, one has $n_\varphi = 1$ in the approach (2.54) and in the eqns. (2.62) to (2.73):

$$F_{n_\varphi}^{TM} = \cos\varphi \quad , \tag{A3.24}$$

$$F_{n_\varphi}^{TE} = \sin\varphi \quad . \tag{A3.25}$$

The equations of the field components in layer "0" are now determined completely. In order to arrive at a closed representation a minor rewriting will be carried out. As is obvious from relationships (A3.13) and (A3.14) one can, in the $z \leq 0$ region, describe the influence of the screen by means of a factor common to all wave amplitudes. Using the wave amplitudes $A_{0\pm}$ and $C_{0\pm}$ from eqns.(A3.22) and (A3.23) one obtains

$$A_0(k_\rho) = S_-^{TM}(k_\rho)\, A_{0-}(k_\rho) \quad \text{and} \quad C_0(k_\rho) = S_-^{TE}(k_\rho)\, C_{0-}(k_\rho) \tag{A3.26}$$

with the abbreviation

$$S_-^{TM/TE}(k_\rho) = \frac{1-/+\Gamma_s^{TM/TE}e^{-j2k_{z0}d_s}}{1 - \Gamma_0^{TM/TE}(k_\rho)\,\Gamma_s^{TM/TE}\,e^{-j2k_{z0}(d_s + d_0)}} \quad . \tag{A3.27}$$

If there is no screen ($d_s \to \infty$) $\Gamma_s^{TM/TE} = 0$ and therefore $S_-^{TM/TE} = 1$. For the region $0 \leq z \leq d_s$ factors $S_+^{TM/TE}$ can be defined in an analogous way. They are determined either by eqns. (A3.17) to (A3.20) or by $S_-^{TM/TE}$ by exchanging the subscripts "s" and "0" of $\Gamma^{TM/TE}$ and d:

$$S_+^{TM/TE}(k_\rho) = \frac{1-/+\Gamma_0^{TM/TE}e^{-j2k_{z0}d_0}}{1 - \Gamma_0^{TM/TE}\,\Gamma_s^{TM/TE}\,e^{-j2k_{z0}(d_0 + d_s)}} \quad . \tag{A3.28}$$

If there is no screen one has

$$S_+^{TM/TE} = 1-/+\Gamma_0^{TM/TE}e^{-j2k_{z0}d_0} \quad . \tag{A3.29}$$

If perfectly conducting material is used for the screen, one has to set $\Gamma_s^{TM/TE} = \pm 1$ according to eqns. (A2.17) and (A2.18).

A3.2 Equations in the region $z < -d_0$

The wave amplitudes for the deeper located layers can be determined using the equations for the field in the "0" layer with the help of eqns. (A2.6) and (A2.7) where the $\tilde{a}_i$, $\tilde{b}_i$, $\tilde{c}_i$ and $\tilde{d}_i$ have been fixed by eqns. (A2.2) to (A2.5). A clear concise presentation of the relationship between the amplitudes of two successive layers is achieved, if one introduces propagation matrices used in [2.78] which are defined by

$$\begin{bmatrix} \tilde{a}_i \\ \tilde{b}_i \end{bmatrix} = \begin{bmatrix} M_i^{i-1} \end{bmatrix} \cdot \begin{bmatrix} \tilde{a}_{i-1} \\ \tilde{b}_{i-1} \end{bmatrix} \quad , \tag{A3.30}$$

$$\begin{bmatrix} \tilde{c}_i \\ \tilde{d}_i \end{bmatrix} = \begin{bmatrix} N_i^{i-1} \end{bmatrix} \cdot \begin{bmatrix} \tilde{c}_{i-1} \\ \tilde{d}_{i-1} \end{bmatrix} \quad . \tag{A3.31}$$

Due to the boundary conditions the propagation matrices read

$$[M_i^{i-1}] = \frac{1}{2} \begin{bmatrix} \varepsilon(+)_i^{i-1} e(-)_{i-1} & \varepsilon(-)_i^{i-1} e(-)_{i-1} \\ \varepsilon(-)_i^{i-1} e(+)_{i-1} & \varepsilon(+)_i^{i-1} e(+)_{i-1} \end{bmatrix} \quad , \tag{A3.32}$$

$$[N_i^{i-1}] = \frac{1}{2} \begin{bmatrix} \mu(+)_i^{i-1} e(-)_{i-1} & \mu(-)_i^{i-1} e(-)_{i-1} \\ \mu(-)_i^{i-1} e(+)_{i-1} & \mu(+)_i^{i-1} e(+)_{i-1} \end{bmatrix} \quad . \tag{A3.33}$$

Apart from the definitions (A2.9) and (A2.10) the abbreviation

$$e(\pm)_i = e^{\pm j k_{z(i+1)}(d_{i+1} - d_i)} \tag{A3.34}$$

was used.

Repeated application of (A3.30) and (A3.31) yields for the TM-amplitudes

$$\begin{bmatrix} \tilde{a}_i \\ \tilde{b}_i \end{bmatrix} = \begin{bmatrix} M_i^{i-1} \end{bmatrix} \cdot \begin{bmatrix} M_{i-1}^{i-2} \end{bmatrix} \cdots \begin{bmatrix} M_1^0 \end{bmatrix} \cdot \begin{bmatrix} \tilde{a}_0 \\ \tilde{b}_0 \end{bmatrix} \tag{A3.35}$$

and for the TE-amplitudes

$$\begin{bmatrix} \tilde{c}_i \\ \tilde{d}_i \end{bmatrix} = \begin{bmatrix} N_i^{i-1} \end{bmatrix} \cdot \begin{bmatrix} N_{i-1}^{i-2} \end{bmatrix} \cdots \begin{bmatrix} N_1^0 \end{bmatrix} \cdot \begin{bmatrix} \tilde{c}_0 \\ \tilde{d}_0 \end{bmatrix} \quad . \tag{A3.36}$$

We shall rewrite eqns. (A3.35) and (A3.36), which are somewhat unhandy because of the i-fold matrix product, by introducing transmission coefficients T_i^{TM} or T_i^{TE} defined by

$$A_i = T_i^{TM} A_0 \quad , \tag{A3.37}$$

$$C_i = T_i^{TE} C_0 \quad . \tag{A3.38}$$

Starting from the reflection coefficients for the TM- and the TE-waves in the interface $z = -d_i$ one obtains e.g. the wave amplitude $\tilde{b}_{i-1}$ of the reflected TM-wave within the layer $i - 1$ from the amplitude of the incident wave $\tilde{a}_{i-1}$ via the relationship

$$\tilde{b}_{i-1} = \tilde{a}_{i-1} \Gamma_{i-1}^{TM} \quad . \tag{A3.39}$$

Therefore, $\tilde{a}_i$ can be obtained directly from $\tilde{a}_{i-1}$ by eqn. (A3.30). Thus

$$\tilde{a}_i = \frac{1}{2} e(-)_{i-1} \left[\varepsilon(+)_i^{i-1} + \varepsilon(-)_i^{i-1} \Gamma_{i-1}^{TM} \right] \tilde{a}_{i-1} \quad . \tag{A3.40}$$

With eqns. (A2.2) to (A2.5) one obtains

$$A_i = \frac{1}{2} e^{-j(k_{z(i-1)} - k_{zi})d_{i-1}} \left[\varepsilon(+)_i^{i-1} + \varepsilon(-)_i^{i-1} \Gamma_{i-1}^{TM} \right] A_{i-1} \quad . \tag{A3.41}$$

If one applies the recursive relation (A3.41) to determine the known amplitude A_0 one is led to two equations for the required transmission coefficients:

$$T_i^{TM}(k_\rho) = \tfrac{1}{2^i} \prod_{q=1}^{i} \left[\varepsilon(+)_q^{q-1} + \varepsilon(-)_q^{q-1} \, \Gamma_{q-1}^{TM}(k_\rho)\right] \, e^{\displaystyle -j\sum_{\nu=1}^{i}(k_{z(\nu-1)} - k_{z\nu})d_{\nu-1}} \, ,$$

$$(A3.42)$$

$$T_i^{TE}(k_\rho) = \tfrac{1}{2^i} \prod_{q=1}^{i} \left[\mu(+)_q^{q-1} + \mu(-)_q^{q-1} \, \Gamma_{q-1}^{TE}(k_\rho)\right] \, e^{\displaystyle -j\sum_{\nu=1}^{i}(k_{z(\nu-1)} - k_{z\nu})d_{\nu-1}} \, .$$

$$(A3.43)$$

Eqn. (A3.43) has been obtained by involving duality, which means that one gets T_i^{TE} from T_i^{TM} by interchanging ε and μ and TE and TM. The still unknown amplitudes B_i, D_i are connected with the amplitudes A_i, C_i, which have been determined by the transmission coefficients, via the reflection coefficients. Hence, the general approach for the components E_z^{TM}, H_z^{TE} (eqns. (2.64) and (2.73)) in the region "i" can be written as follows

$$E_{zi}^{TM} = \int\limits_{-\infty}^{+\infty} A_0(k_\rho) T_i^{TM}(k_\rho) \left[e^{jk_{zi}z} + \Gamma_i^{TM}(k_\rho) e^{-j2k_{zi}d_i} \quad e^{-jk_{zi}z}\right]$$

$$\cdot H_1^{(2)}(k_\rho\rho) F_1^{TM}(\varphi) \, dk_\rho \quad , \qquad (A3.44)$$

$$H_{zi}^{TE} = \int\limits_{-\infty}^{+\infty} C_0(k_\rho) T_i^{TE}(k_\rho) \left[e^{jk_{zi}z} + \Gamma_i^{TE}(k_\rho) e^{-j2k_{zi}d_i} e^{-jk_{zi}z}\right]$$

$$\cdot H_1^{(2)}(k_\rho\rho) F_1^{TE}(\varphi) \, dk_\rho \quad . \qquad (A3.45)$$

With the definition

$$T_0^{TM/TE} = 1 \qquad (A3.46)$$

this representation is also valid for the region "0" ($-d_0 \leq z \leq 0$).

We now make use of the amplitudes A_0 and C_0 which are known from section A3.1 (eqns. (A3.13) and (A3.14)) and the cylinder and trigonometric functions specified by eqns. (A3.24) and (A3.25). Based on the general approach (eqns. (2.62) to (2.73)) we then can formulate an integral representation which is valid in the entire region $z \leq 0$. In order to stress that the equations are valid denoting the field of a point source the terms for the field components are preceded by a "d", according to the definition (2.8):

$$dE_{\rho i}^{TM} = A_E \frac{1}{k\varepsilon_{r0}} \int\limits_{-\infty}^{\infty} -k_{zi}k_\rho\, S_-^{TM}(k_\rho)\, T_i^{TM}(k_\rho) \tag{A3.47}$$

$$\left[e^{+jk_{zi}z} - \Gamma_i^{TM}(k_\rho)\, e^{-jk_{zi}(z+2d_i)}\right] H_1^{(2)'}(k_\rho\rho)\cos\varphi\; dk_\rho \quad,$$

$$dE_{\varphi i}^{TM} = A_E \frac{1}{k\varepsilon_{r0}} \int\limits_{-\infty}^{\infty} \frac{k_{zi}}{\rho}\, S_-^{TM}(k_\rho)\, T_i^{TM}(k_\rho) \tag{A3.48}$$

$$\left[e^{+jk_{zi}z} - \Gamma_i^{TM}(k_\rho)\, e^{-jk_{zi}(z+2d_i)}\right] H_1^{(2)}(k_\rho\rho)\sin\varphi\; dk_\rho \quad,$$

$$dE_{zi}^{TM} = A_E \frac{1}{k\varepsilon_{r0}} \int\limits_{-\infty}^{\infty} -jk_\rho^2\, S_-^{TM}(k_\rho)\, T_i^{TM}(k_\rho) \tag{A3.49}$$

$$\left[-e^{+jk_{zi}z} - \Gamma_i^{TM}(k_\rho)\, e^{-jk_{zi}(z+2d_i)}\right] H_1^{(2)}(k_\rho\rho)\cos\varphi\; dk_\rho \quad,$$

$$dH_{\rho i}^{TM} = A_H \frac{\varepsilon_{ri}}{\varepsilon_{r0}} \int\limits_{-\infty}^{\infty} -\frac{1}{\rho}\, S_-^{TM}(k_\rho)\, T_i^{TM}(k_\rho) \tag{A3.50}$$

$$\left[-e^{+jk_{zi}z} - \Gamma_i^{TM}(k_\rho)\, e^{-jk_{zi}(z+2d_i)}\right] H_1^{(2)}(k_\rho\rho)\sin\varphi\; dk_\rho \quad,$$

$$dH_{\varphi i}^{TM} = A_H \frac{\varepsilon_{ri}}{\varepsilon_{r0}} \int\limits_{-\infty}^{\infty} -k_\rho\, S_-^{TM}(k_\rho)\, T_i^{TM}(k_\rho) \tag{A3.51}$$

$$\left[-e^{+jk_{zi}z} - \Gamma_i^{TM}(k_\rho)\, e^{-jk_{zi}(z+2d_i)}\right] H_1^{(2)'}(k_\rho\rho)\cos\varphi\; dk_\rho \quad,$$

$$dH_{zi}^{TM} = 0 \quad, \tag{A3.52}$$

$$dE_{\rho i}^{TE} = A_E k\mu_{ri} \int\limits_{-\infty}^{\infty} -\frac{1}{k_{z0}\rho}\, S_-^{TE}(k_\rho)\, T_i^{TE}(k_\rho) \tag{A3.53}$$

$$\left[e^{+jk_{zi}z} + \Gamma_i^{TE}(k_\rho)\, e^{-jk_{zi}(z+2d_i)}\right] H_1^{(2)}(k_\rho\rho)\cos\varphi\; dk_\rho \quad,$$

$$dE_{\varphi i}^{TE} = A_E k\mu_{ri} \int\limits_{-\infty}^{\infty} \frac{k_\rho}{k_{z0}}\, S_-^{TE}(k_\rho)\, T_i^{TE}(k_\rho) \tag{A3.54}$$

$$\left[e^{+jk_{zi}z} + \Gamma_i^{TE}(k_\rho)\, e^{-jk_{zi}(z+2d_i)}\right] H_1^{(2)'}(k_\rho\rho)\sin\varphi\; dk_\rho \quad,$$

$$dE_{zi}^{TE} = 0 \qquad , \qquad\qquad\qquad\qquad\qquad (A3.55)$$

$$dH_{\rho i}^{TE} = A_H \int\limits_{-\infty}^{\infty} -k_\rho \frac{k_{zi}}{k_{z0}} S_-^{TE}(k_\rho) T_i^{TE}(k_\rho) \qquad\qquad (A3.56)$$

$$\left[-e^{+jk_{zi}z} + \Gamma_i^{TE}(k_\rho) e^{-jk_{zi}(z+2d_i)} \right] H_1^{(2)'}(k_\rho\rho) \sin\varphi \; dk_\rho \quad ,$$

$$dH_{\varphi i}^{TE} = A_H \int\limits_{-\infty}^{\infty} -\frac{k_{zi}}{\rho k_{z0}} S_-^{TE}(k_\rho) T_i^{TE}(k_\rho) \qquad\qquad (A3.57)$$

$$\left[-e^{+jk_{zi}z} + \Gamma_i^{TE}(k_\rho) e^{-jk_{zi}(z+2d_i)} \right] H_1^{(2)}(k_\rho\rho) \cos\varphi \; dk_\rho \quad ,$$

$$dH_{zi}^{TE} = A_H \int\limits_{-\infty}^{\infty} -j\frac{k_\rho^2}{k_{z0}} S_-^{TE}(k_\rho) T_i^{TE}(k_\rho) \qquad\qquad (A3.58)$$

$$\left[e^{+jk_{zi}z} + \Gamma_i^{TE}(k_\rho) e^{-jk_{zi}(z+2d_i)} \right] H_1^{(2)}(k_\rho\rho) \sin\varphi \; dk_\rho \quad .$$

The abbreviations A_H, A_E denote the amplitudes of the magnetic and electric field components, respectively, which are proportional to the dipole moment:

$$A_H = \frac{I_x dl_x}{8\pi} \; , \qquad A_E = \frac{I_x dl_x}{8\pi} \sqrt{\frac{\mu_0}{\varepsilon_0}} \quad . \qquad (A3.59)$$

All equations for the region $0 \leq z \leq d_s$ can be derived in an analogous way. We only give the equations for the E_{z0}^{TM} and the H_{z0}^{TE} components:

$$dE_{z0}^{TM} = A_E \frac{1}{k\varepsilon_{r0}} \int\limits_{-\infty}^{\infty} jk_\rho^2 S_+^{TM}(k_\rho) \left[-e^{-jk_{z0}z} - \Gamma_s^{TM} e^{-jk_{z0}(-z+2d_s)} \right]$$

$$\cdot H_1^{(2)}(k_\rho\rho) \cos\varphi \; dk_\rho \quad , \qquad\qquad (A3.60)$$

$$dH_{z0}^{TE} = A_H \int\limits_{-\infty}^{\infty} -j\frac{k_\rho^2}{k_{z0}} S_+^{TE}(k_\rho) \left[+e^{-jk_{z0}z} + \Gamma_s^{TE} e^{-jk_{z0}(-z+2d_s)} \right]$$

$$\cdot H_1^{(2)}(k_\rho\rho) \sin\varphi \; dk_\rho \quad . \qquad\qquad (A3.61)$$

Comparison with eqns. (A3.47) to (A3.58) for $i = 0$ reveals that principally S_- has to be replaced by S_+, Γ_0 by Γ_s and d_0 by d_s. Apart from this, a few signs are altered because, for instance, the part radiated by the source propagates in positive z-direction for $z > 0$, and in negative z-direction for $z < 0$.

Because of the similarity of the integrands of all 10 field components a general representation can be found, that is valid in the entire region $z \leq d_s$:

$$d\Psi_{\xi\pm}^{TM/TE} = AM_{\Psi\pm}^{TM/TE} \int_{-\infty}^{\infty} f_{\Psi\xi}^{TM/TE}(k_\rho\rho)\, S_\pm^{TM/TE}(k_\rho)\, T_i^{TM/TE}(k_\rho)\, \{I\}$$

$$Z_{\Psi\xi}^{TM/TE}(k_\rho\rho)\, F_{\Psi\xi}(\varphi)\, dk_\rho \qquad (A3.62)$$

with

$$\{I\} = \left\{ s_{\Psi\xi}\, e^{+j(S_-^+)k_{zi}z} - / + \Gamma^{TM/TE}(k_\rho)\, e^{-j(S_-^+)k_{zi}z} \quad e^{-jk_{z\pm}} \right\} . \qquad (A3.63)$$

The abbreviations which have been used are given below:

$$\left. \begin{array}{l} \Psi = E, H \\ \xi = \rho, \varphi, z \end{array} \right\} \quad \begin{array}{c} + \\ - \end{array} \right\} \text{for} \left\{ \begin{array}{c} z \geq 0 \\ z \leq 0 \end{array} \right., \qquad (A3.64)$$

$$AM_{E-}^{TM} = \frac{A}{\omega\varepsilon_0\varepsilon_{r0}}, \qquad AM_{E+}^{TM} = \frac{A}{\omega\varepsilon_0\varepsilon_{r0}}, \qquad (A3.65)$$

$$AM_{E-}^{TE} = A\omega\mu_0\mu_{ri}, \qquad AM_{E+}^{TE} = A\omega\mu_0\mu_{r0}, \qquad (A3.66)$$

$$AM_{H-}^{TM} = A\frac{\varepsilon_{ri}}{\varepsilon_{r0}}, \qquad AM_{H+}^{TM} = A, \qquad (A3.67)$$

$$AM_{H-}^{TE} = A, \qquad AM_{H+}^{TE} = A, \qquad (A3.68)$$

$$A = \frac{I_x dl_x}{8\pi}, \qquad (A3.69)$$

ξ	ρ		φ		z	
	TM	TE	TM	TE	TM	TE
$f_{E_{\xi+}}(k_\rho\rho)$	$-k_{z0}k_\rho$	$-\dfrac{1}{k_{z0}\rho}$	$\dfrac{k_{z0}}{\rho}$	$\dfrac{k_\rho}{k_{z0}}$	jk_ρ^2	$--$
$f_{E_{\xi-}}(k_\rho\rho)$	$-k_{zi}k_\rho$	$-\dfrac{1}{k_{z0}\rho}$	$\dfrac{k_{zi}}{\rho}$	$\dfrac{k_\rho}{k_{z0}}$	$-jk_\rho^2$	$--$
$f_{H_{\xi+}}(k_\rho\rho)$	$\dfrac{1}{\rho}$	k_ρ	k_ρ	$\dfrac{1}{\rho}$	$--$	$-\dfrac{jk_\rho^2}{k_{z0}}$
$f_{H_{\xi-}}(k_\rho\rho)$	$-\dfrac{1}{\rho}$	$-\dfrac{k_\rho k_{zi}}{k_{z0}}$	$-k_\rho$	$-\dfrac{k_{zi}}{\rho k_{z0}}$	$--$	$-\dfrac{jk_\rho^2}{k_{z0}}$

$$(A3.70)$$

$$S_\pm^{TM/TE} = \left\{ \begin{array}{ll} S_+^{TM/TE} & \text{for } z \geq 0 \\ S_-^{TM/TE} & \text{for } z \leq 0 \end{array} \right., \qquad (A3.71)$$

ξ	ρ	φ	z
s_{E_ξ}	$+1$	$+1$	-1
s_{H_ξ}	-1	-1	$+1$

$$(A3.72)$$

$$S_-^+ = \begin{cases} -1 & \text{for } z \geq 0 \\ +1 & \text{for } z \leq 0 \end{cases} \quad, \tag{A3.73}$$

$$\Gamma^{TM/TE} = \begin{cases} \Gamma_s^{TM/TE} & \text{for } z \geq 0 \\ \Gamma_i^{TM/TE} & \text{for } z \leq 0 \end{cases} \quad, \tag{A3.74}$$

$$k_{z\pm} = \begin{cases} 2k_{z0}d_s & \text{for } z \geq 0 \\ 2k_{zi}d_i & \text{for } z \leq 0 \end{cases} \quad, \tag{A3.75}$$

ξ	ρ		φ		z	
	TM	TE	TM	TE	TM	TE
Z_{E_ξ}	$H_1^{(2)'}$	$H_1^{(2)}$	$H_1^{(2)}$	$H_1^{(2)'}$	$H_1^{(2)}$	$--$
Z_{H_ξ}	$H_1^{(2)}$	$H_1^{(2)'}$	$H_1^{(2)'}$	$H_1^{(2)}$	$--$	$H_1^{(2)}$

$$\tag{A3.76}$$

ξ	ρ	φ	z
F_{E_ξ}	$\cos\varphi$	$\sin\varphi$	$\cos\varphi$
F_{H_ξ}	$\sin\varphi$	$\cos\varphi$	$\sin\varphi$

$$\tag{A3.77}$$

List of symbols

a	Radius of a wire	(2.23)
$\tilde{a}_i$	Amplitudes of the $TM-$ waves travelling in the direction $-z$ in the i-th layer related to the plane $z = -d_i$	(A2.2)
a_{0-3}	Abbreviation, introduced in eqn. (3.84) and (3.85)	
A, A_i	Amplitudes of the $TM-$ waves travelling in the direction $-z$ in the i-th layer related to the plane $z = 0$	(2.54), (A2.2)
$A_{0\pm}$	Amplitudes of the $TM-$ waves excited by a point source, which are travelling in the direction $\pm z$	(A3.1)
A_r	Amplitudes of the $TM-$ waves in the "0"– layer reflected by a perfect screen	(A3.1)
A_E, A_H	Factors proportional to the amplitudes of the dipole moment	(A3.59)
$AM_{\psi\pm}^{TM/TE}$	Factors proportional to the amplitudes of the dipole moment	(A3.62)
$\tilde{b}_i$	Amplitudes of the $TM-$ waves travelling in the direction $-z$ in the i-th layer related to the plane $z = -d_i$	(A2.4)
B, B_i	Amplitudes of the $TM-$ waves travelling in the direction $+z$ in the i-th layer related to the plane $z = 0$	(2.54), (A2.4)
B_{pq}	Imaginary part of Y_{pq}	(Fig. 3.25)
B_r	Amplitudes of the $TM-$ waves reflected by the interface $z = -d_0$ in the "0"– layer	(A3.5)

B_{0p} Factor proportional to the dipole moment in the equation for the amplitudes of the $TM-$ surface waves   (3.44)

$b_{g\varrho,\varphi}$ 3–dB beam width of the mainlobe of $|E_\varrho/\cos\varphi|$, $|E_\varphi/\sin\varphi|$ at the interface fat– muscle tissue   (4.11), (4,12)

$b_{fx,y}$ 3–dB beam width of the mainlobe in the $xz-,yz-$ plane at $z = z_f$ with z_f equal to the depth of the focus   (section 4.3.4.1)

c Velocity of light in free space   (6.31)

$\tilde{c}_i$ Amplitudes of the $TM-$ waves travelling in the direction $-z$ in the i–th layer related to the plane $z = -d_i$ (A2.3)

C Middle axis of the strip   (Fig. 2.4)

C, C_i Amplitudes of the $TE-$ waves travelling in the direction $-z$ in the i–th layer related to the plane $z = 0$ (2.68), (A2.3)

C_r Amplitude of the $TE-$ waves in the "0"– layer reflected by a perfect screen   (A3.2)

C^+ Middle axis of the feed area   (Fig. 2.4)

$C_{0\pm}$ Amplitudes of the $TE-$ waves excited by a point source, which are travelling in the direction $\pm z$ (A3.2)

$d_i, d_{\hat{i}}$ Spacing of the interface $i, \hat{i}$ to the origin of the coordinate system   (Fig.2.1)

$\tilde{d}_i$ Amplitudes of the $TM-$ waves travelling in the direction $-z$ in the i–th layer related to the plane $z = -d_i$ (A2.5)

d_s Spacing of the perfect screen to the origin of the coordinate system   (Fig.2.7)

$d\psi_{\xi\pm}^{TM/TE}$ Abbreviation, introduced in eqn.(A3.62)

D, D_i Amplitudes of the $TE-$ waves travelling in the direction $+z$ in the i–th layer related to the plane $z = 0$ (2.68), (A2.5)

D_f Difference between the level of $|\vec{E}|$ in the focus and the maximum level in the skin layer   (4.25)

$\tilde{D}_f$	Difference between the level of $|\tilde{E}|$ in the focus and the maximum level in the skin layer	(4.30)
D_r	Amplitudes of the $TE-$ waves reflected by the interface $z = -d_0$ in the "0"– layer	(A3.4)
D_{0p}	Factor proportional to the dipole moment in the equation for the amplitudes of the $TE-$ surface waves	(3.45)
$D_{02\varrho,\varphi}$	Difference between the level of $E_{\varrho,\varphi}$ in the interface fat– muscle tissue and that at the surface of the skin	(4.8)

e_q	Abbreviation, introduced in eqn. (6.66)	
$e(\pm)_i$	Abbreviation, introduced in eqn. (A.3.34)	
$\vec{e}(\vec{r},t)$	Electrical field strength in the time domain	(6.2)
$\vec{E}$	Electrical field strength	(2.14)
$|\tilde{E}|$	Power absorbed per volume	(section 4.3.4.2)
$\vec{E}_i$	Impressed field	(Fig. 2.3)
$\vec{E}^{TM/TE}$	Electrical field strength of the TM/TE part of the field	(2.52), (2.56)
$d\vec{E}$	Electrical field of a point source	(2.8)
$d\vec{E}_i^{TM/TE}$	Electrical field strength of the TM/TE part of the field excited by a point source in the i–th layer	(A3.47)
$d\vec{E}_{i\infty}^{TM/TE}$	Electrical field strength of the TM/TE part of the field excited by a point source for very small distances between source and observation point in the i–th layer	(2.143)
$d\vec{E}_s^{TM/TE}$	Electrical field strength of the TM/TE part of the field excited by a point source in the farfield (saddle-point solution)	(2.100)
$d\vec{E}_{iw}^{TM/TE}$	Electrical field strength of a $TM/TE-$ surface wave excited by a point source in the i–th layer	(3.44)
$d\hat{E}_{\varrho,\varphi 0w}^{TM/TE}$	Amplitudes of a $TM/TE-$ surface wave in the farfield	(3.136)
$d\vec{E}(\vec{r})$	Electrical field strength of a point source	(2.8)

f	Frequency	(Fig. 5.20)
$f(w)$	Abbreviation, introduced in eqn. (2.86)	

$f_{cn}^{TM/TE}$	Cut–off frequency of the n–th TM/TE– surface wave	(3.29)
f_m	Centre frequency of a Gaussian pulse	(6.3)
$f_{max,min}$	Upper, lower limit of a Gaussian pulse	(6.5)
$f_p(\vec{r})$	Basis function	(A1.2)
f_{tr}	Frequency of a model	(5.21)
$f_{NR}^{TM/TE}$	Zeros of the real part of $G_{\vartheta,\varphi}^{\infty}$	(6.43)
$f_{NI}^{TM/TE}$	Zeros of the imaginary part of $G_{\vartheta,\varphi}^{\infty}$	(6.44)
$f_{\psi\xi}^{TM/TE}$	Abbreviation, introduced in eqn. (A3.62)	
$f_{\vartheta,\varphi}(\vartheta)$	$\vartheta,\varphi-$ dependency of the space wave excited by a point source	(3.63), (3.64)
$f_{\infty}(k_\varrho)$	$f(k_\varrho)$ for $k_\varrho \gg k$	(2.138)
F	Abbreviation, introduced in eqn. (A1.1)	
$\mathcal{F}^{-1}$	Inverse Fourier transform	(6.50)
$F(\vartheta,\varphi)$	Integral over a current distribution of a strip for the determination of the radiation pattern	(3.96)
$F_{\psi\xi}(\varphi)$	Abbreviation, introduced in eqn. (A3.62)	
$F_{n_\varphi},\ F_{n_\varphi}^{TM/TE}$	φ–dependency of the ansatz for the field on φ	(2.35)
$g(x,x')$	Abbreviation, introduced in (2.22)	
$g(k_\varrho),\ g(k_0\sin w)$	Abbreviation, introduced in eqn. (2.77),(2.84)	
$g_{\vartheta,\varphi}(t)$	Inverse Fourier transform of $G_{\vartheta,\varphi}$	(6.58), (6.59)
$g_{\vartheta,\varphi}(\vartheta,d_i)$	Abbreviation, introduced in eqn. (3.113)	
$\overleftrightarrow{G}$	Green's tensor	(2.8)
G_{pq}	Real part of Y_{pq}	(Fig. 3.25)
$G_{\varphi,\vartheta}$	Components of the Green's tensor in the farfield	(6.6)
$G_{\varphi,\vartheta}^{\infty}$	Abbreviation, introduced in eqn. (6.6)	
$h_{g1}(t)$	Abbreviation, introduced in eqn. (6.50)	
$h_{\vartheta,\varphi}(\vartheta,d_i)$	Abbreviation, introduced in eqn. (3.113)	
$\vec{H}$	Magnetic field strength	(2.12)
$\vec{H}^{TM/TE}$	Magnetic field strength of the TM/TE part of the field	(2.51), (2.58)
$d\vec{H}_i^{TM/TE}$	Magnetic field strength of the TM/TE part of the field excited by a point source in the i–th layer	
$H_{n_\varphi}^{(1/2)}(k_\varrho\varrho)$	Hankel functions of the 1st/2nd kind and the order n_φ	(2.55)

$H_{1/2}(f)$	Abbreviation, introduced in eqn. (6.8)	
I	Abbreviation, introduced in eqn. (2.98)	
I	Current	(Fig. 3.18)
$\vec{I}(\vec{r})dl$	Current element of the length l	(2.6)
I_s	Current on the strip	(2.20)
$\hat{I}_p$	Coefficients of the series expansion of I_s	(A1.2)
$I_{1,2}(0)$	Current at the gate 1, 2	(Fig. 3.22)
$\{I\}$	Abbreviation, introduced in eqn. (A3.63)	
$\{I\}_\infty$	$\{I\}$ for $k_\varrho \gg k$	(2.139)
$\vec{J}$	Current density	(2.12)
$\vec{J}_F$	Surface current	(2.1)
$J_{n_\varphi}(k_\varrho \varrho)$	Bessel function of the 1st kind and the order n_φ	(2.34)
$k,\ k_e,\ k_i$	Propagation constant of a uniform plane wave in a space with $\varepsilon_r = \mu_r = 1$; $\varepsilon_{re},\ \mu_{re} = 1$; $\varepsilon_{ri},\ \mu_{ri}$	(2.3),(3.98),(2.5)
k_{zi}	$z-$ component of the propagation vector in the i-th layer	(2.36)
$k_{z\pm}$	Abbreviation, introduced in eqn. (A3.63)	
k_ϱ	Variable of integration	(2.34)
$k_{\varrho b_{1,2}}$	Location of a branch point in the $k_\varrho-$ plane	(section 2.41)
$k_{\varrho p}^{TM/TE}$	Location of a pole in the $k_\varrho-$ plane	
$k_{\varrho \varepsilon}$	Lower limit for an integration	(2.140)
k_z^*	Equal to $j k_z$	(3.37)
$\Delta k'_{\varrho p}$	Surrounding of a pole on the real $k_\varrho-$ axis (Fig. 2.11)	
l	Half length of a dipole	(Fig. 3.17)
$l_{res1,2}$	1st, 2nd resonance length of a printed dipole	(3.88)
L	Linear operator	(A.1.1)
$\hat{m}$	Number of layers above the source	(Fig. 2.1)
M	Abbreviation, introduced in eqn. (3.169)	

$[M_i^{i-1}]$	Propagation matrices	(A3.30)
n	Number of layers below the source	(Fig. 2.1)
n_φ	Constant of separation	(2.37)
$[N_i^{i-1}]$	Propagation matrices	(A3.31)
P_{in}	Input power	(3.164)
P_v	Power converted to heat in a volume V	(4.1)
P_{sr}	Total radiated power	(3.103)
P_{sw}	Power carried by the TM_1- surface wave	(3.106)
$P_{STI}^{TM/TE}$	Abbreviation, introduced in eqn. (3.17)	
q	Abbreviation, introduced in eqn. (2.86)	
$Q_{1-3}^{TM/TE}$	Abbreviation, introduced in eqn. (2.125)	
$Q_{2\infty}^{TM/TE}$	Abbreviation, introduced in eqn. (2.133)	
$r,\ \vartheta,\ \varphi$	Spherical coordinates	
r'	Abbreviation, introduced in eqn. (2.83)	
r_i	Reflection coeffient	(5.26)
$r_i^{TM/TE}$	Reflection coeffient	(6.34)
R	Abbreviation, introduced in eqn. (2.86)	
R_g	Internal resistance of a generator	(6.1)
$\bar{R}_{in}$	Arithmetic mean of the reflection coefficients	(3.165)
$R_n(\vartheta,\varphi)$	Reflection coefficient of the n-th element of an array	(3.163)
$R_{0,s}^{TM/TE}$	Abbreviation, introduced in eqn. (A.3.21)	
R_{pq}	Real part of Z_{pq}	(Fig. 5.11)
$R_{\psi 1-5}$	Abbreviation, introduced in eqn. (2.117)	
s	tangent of C	(Fig. 2.4)
$s_{\psi\xi},\ S\pm$	Abbreviations, introduced in (A3.63)	
S	Surface of the strips	(Fig. 2.4)
S^+	Surface of the feed area	(Fig. 2.4)
S_z	z-component of the power density carried by a uniform plane wave	(Fig. 4.6)

$S_{\pm}^{TM/TE}$	Term which takes into account the perfect screen above the source	(A3.26)
$T_i^{TM/TE}$	Transmission coefficients	(A3.37)
$u(t)$	open–circuit voltage of the source in the time domain	(6.1)
$\vec{u}_{\varrho,\varphi,z}$, $\vec{u}_{x,y,z}$	Unit vectors	(2.60),(2.61)
$U(x)$	Unit step function	(2.98)
$U_{1,2}(0)$	Voltage at the gate 1, 2	(Fig. 3.22)
$U_g(f)$	open–circuit voltage of the source	(6.1)
$v_{ph}^{TM/TE}$	Phase velocity of a TM/TE– surface wave	(3.54)
w	Width of the strips	(2.16)
w	Complex variable	(2.79)
$w_{p,s,v}$	Location of a pole, a saddle point, a branch point in the w– plane	(2.92),(2.95)
w_q	Weighting function	(A1.4)
W	Sommerfeld path of integration	(2.84)
W_s	Path of steepest descent	(Fig. 2.9)
x, y, z	Rectangular coordinates	(2.6)
Δx, Δy	Spacing between the dipoles of an array	(Fig. 5.7)
X_{pq}	Imaginary part of Z_{pq}	(Fig. 5.11)
Y_{pq}	Mutual coupling between the dipoles p and q, admittance	(3.133)
z'	Abbreviation, introduced in eqn. (2.77)	
Z_e	Intrinsic impedance of a space with ε_{re}	(6.12)
z_f	Depth of the focus	(4.2)
Z_{in}	Input impedance of a dipole	(Fig. 3.20)
Z_{pq}	Mutual coupling between the dipoles p and q, impedance	(3.132)
$Z_{\psi\xi}^{TM/TE}$	Abbreviation, introduced in eqn. (A3.62)	
$\alpha_{e,h,t,r}$	Scale factors	(5.9)

$\gamma_0^{TM/TE}$	Inverse Fourier transform of $\Gamma_0^{TM/TE}$	(6.58)
$\delta(t)$	Dirac impulse	(6.58)
$\varepsilon(\pm)_q^p$	Abbreviation, introduced in eqn. (A2.9)	
ε_0	Permittivity of free space	(2.3)
ε_{re}	Arithmetic mean of ε_0 and ε_{r1}	(3.99)
$\varepsilon_{r\,eff}$	Effective ε_r	(3.41)
ε_{r_i}	Relative permittivity of the i–th layer	(2.3)
ε_{tr}	Relative permittivity of a model	(5.13)
$\eta_{0,1}$	Abbreviation, introduced in eqn. (3.67)	
$\vartheta_{bl}, \varphi_{bl}$	Blind angle	(3.162)
κ_{tr}	Conductivity of a model	(5.15)
$\lambda, \lambda_i, \lambda_e$	Wavelength of a uniform plane wave in free space, in homogeneous space with $\varepsilon_{r_{i,e}}, \mu_{r_{i,e}}$ (section 2.2.1),(3.98)	
$\lambda_{cn}^{TM/TE}$	Cut–off wavelength of the n–th TM/TE– surface wave (Fig. 3.9)	
λ_g	Guide wavelength (section 3.1)	
$\mu(\pm)_q^p$	Abbreviation, introduced in eqn. (A2.10)	
μ_0	Permeability of free space	(2.3)
μ_{ri}	Relative permeability of the i–th layer	(2.3)
μ_{tr}	Relative permeability of a model	(5.14)
$\psi^{TM/TE}$	z–component of the vector potential of the TM/TE field (2.51),(2.52)	
ϱ, φ, z	Coordinates of a cylindrical coordinate system (2.34)	
σ	Constant specifying the Gaussian pulse	(6.3)
σ	Standard deviation from the arithmetic means (3.161)	
τ_d, τ_l	Delay time	(6.63),(6.51)
$\Delta\tau_p$	Pulse length of the response	(section 6.4.1)
$\Delta\tau_{pe}$	Pulse length of the excitation	(6.53)
$<\varphi, \psi>$	Inner product of the two functions φ and ψ (A1.4)	
ω	Angular frequency	(2.3)
$\Gamma_{i,s}^{TM/TE}$	Reflection coefficients of the TM/TE– waves at the interface $z = -d_i$, $z = +d_s$ (A2.12)	
$\vec{\Pi}, \vec{\Pi}_i$	Hertzian vector, in the i–layer (section 2.3.2)	

REFERENCES

References chapter 2

[2.1] Franz, W., Zur Formulierung des Huygensschen Prinzips. Zeitschrift für Naturforschung 3a [1948], 500-506.

[2.2] Severin, H., Zur Theorie der Beugung elektromagnetischer Wellen. Zeitschrift für Physik 129 [1951], 426-439.

[2.3] Richmond, J.H., A wire-grid model for scattering by conducting bodies. Transact. IEEE AP-14 [1966], 782-786.

[2.4] Denlinger, E.J., A frequency dependent solution for microstrip transmission lines. Transact. IEEE MTT-19 [1971], 30-39.

[2.5] Pocklington, H.C., Electrical oscillations in wires. Proc. Camb. Phil. Soc. 9 [1897], 324-332.

[2.6] Richmond, J.H., Digital computer solutions of the rigorous equations for scattering problems. Proc. IEEE 53 [1965], 796-804.

[2.7] Hallen, E., Theoretical investigations into the transmitting and receiving qualities of antennae. N.A.R. Soc. Sci. Upsal. 11 [1938], 1-44.

[2.8] Borgnis, F.E. und Papas, C.H., Randwertprobleme der Mikrowellenphysik. Springer-Verlag, Berlin 1955, 217-221.

[2.9] King, R.W.P. and Harrison, C.W., Antennas and waves: A modern approach. The M.I.T. Press, Cambridge 1969, 137-248.

[2.10] Chang, S.-K. and Mei, K.K., Multipole expansion technique for electromagnetic scattering by buried objects. Electromagnetics 1 [1981], 73-89.

[2.11] Karlsson, A. and Kristensson, G., Electromagnetic scattering from subterranean obstacles in a stratified ground. Radio Sci. 18 [1983], 345-356.

[2.12] Degauque, P., Nguyen, Q.T. and Cauterman, M., Simulation of the high-frequency response of a heterogeneous ground through the finite difference technique. Transact. IEEE GE-22 [1984], 368-374.

[2.13] Butler, C.M., Current induced on a conducting strip which resides on the planar interface between two semi-infinite half-spaces. Transact. IEEE

AP-32 [1984], 226-231.

[2.14] Sommerfeld, A., Über die Ausbreitung der Wellen in der drahtlosen Telegraphie. Ann. Phys. 28 [1909], 665-736.

[2.15] v.Hörschelmann, H., Über die Wirkungsweise des geknickten Marconischen Senders in der drahtlosen Telegraphie. Jahrb. draht. Telegraphie Telephonie 5 [1912], 14-34 and 188-211.

[2.16] Sommerfeld, A., Über die Ausbreitung der Wellen in der drahtlosen Telegraphie. Ann. Phys. 81 [1926], 1135-1153.

[2.17] Mohsen, A., On the evaluation of Sommerfeld integrals. Proc. IEE Pt. H. 129 [1982], 177-182.

[2.18] Zenneck, J., Über die Fortpflanzung ebener elektromagnetischer Wellen längs einer ebenen Leiterfläche und ihre Beziehung zur drahtlosen Telegraphie. Ann. Phys. 23 [1907], 846-866.

[2.19] Weyl, H., Ausbreitung elektromagnetischer Wellen über einem ebenen Leiter. Ann. Phys. 60 [1919], 481-500.

[2.20] Sommerfeld, A., Über die Ausbreitung der Wellen in der drahtlosen Telegraphie. Ann. Phys. 62 [1920], 95-96.

[2.21] Van der Pol, B. and Niessen, K.F., Über die Ausbreitung elektromagnetischer Wellen über eine ebene Erde. Ann. Phys. 6 [1930], 273-294.

[2.22] Van der Pol, B. and Niessen, K.F., Über die Raumwellen von einem vertikalen Dipolsender auf ebener Erde. Ann. Phys. 10 [1931], 485-510.

[2.23] Van der Pol, B., Über die Ausbreitung elektromagnetischer Wellen. Z. Hochfrequenztechnik 37 [1931], 152-156.

[2.24] Niessen, K.F., Über die entfernten Raumwellen eines vertikalen Dipolsenders oberhalb einer ebenen Erde von beliebiger Dielektrizitätskonstante und beliebiger Leitfähigkeit. Ann. Phys. 18 [1933], 893-912.

[2.25] Van der Pol, B., Theory of the reflection of the light from a point source by a finitely conducting flat mirror, with an application to radiotelegraphy. Physica 2 [1935], 843-853.

[2.26] Niessen, K.F., Eine Verschärfung der verbesserten Sommerfeldschen Fortpflanzungsformel für drahtlose Wellen zur Ausbreitung ihres Gültigkeitsgebietes nach kleineren Abständen. Ann. Phys. 29 [1937], 569-584.

[2.27] Niessen, K.F., Zur Entscheidung zwischen den beiden Sommerfeldschen Formeln für die Fortpflanzung von drahtlosen Wellen. Ann. Phys. 29 [1937], 585-596.

[2.28] Strutt, M.J.O., Strahlung von Antennen unter dem Einfluß der Erdbodeneigenschaften. Ann. Phys. 1 [1929], 721-750.

[2.29] Wise, W.H., Asymptotic dipole radiation formulas. Bell Syst. tech. J. 8 [1929], 662-671.

[2.30] Wise, W.H., The grounded condenser antenna radiation formula. Proc. IRE 19 [1931], 1684-1689.

[2.31] Rice, S.O., Series for the wave function of a radiating dipole at the earth's surface. Bell Syst. tech. J. 16 [1937], 101-109.

[2.32] Fock, V., Zur Berechnung des elektromagnetischen Wechselstromfeldes bei ebener Begrenzung. Ann. Phys. 17 [1933], 401-420.

[2.33] Norton, K.A., Propagation of radio waves over a plane earth. Nature 135 [1935], 954-955.

[2.34] Norton, K.A., The propagation of radio waves over the surface of the earth and in the upper atmosphere (Part I). Proc. IRE 24 [1936], 1367-1387.

[2.35] Norton, K.A., Space and surface waves in radio propagation. Phys. Rev. 52 [1937], 132-133.

[2.36] Norton, K.A., The propagation of radio waves over the surface of the earth and in the upper atmosphere (Part II). Proc. IRE 25 [1937], 1203-1236.

[2.37] Wise, W.H., The physical reality of Zenneck's surface wave. Bell Syst. tech. J. 16 [1937], 35-44.

[2.38] Burrows, C.R., Radio propagation over plane earth – field strength curves. Bell Syst. tech. J. 16 [1937], 45-75.

[2.39] Burrows, C.R., The surface wave in radio propagation over plane earth. Proc. IRE 25 [1937], 219-229.

[2.40] Epstein, P.S., Radio-wave propagation and electromagnetic surface waves. Proc. Nat. Acad. Sci. 33 [1947], 195-199.

[2.41] Kahan, T. and Eckart, G., On the electromagnetic surface wave of Sommerfeld. Phys. Rev. 76 [1949], 406-410.

[2.42] Bouwkamp, C.J., Math. Rev. 9 [1948], 637.

[2.43] Bouwkamp, C.J., On Sommerfeld's surface wave. Phys. Rev. 80 [1950], 294.

[2.44] Ott, H., Die Bodenwelle eines Senders. Z. angew. Phys. 3 [1951], 123-134.

[2.45] Ott, H., Gibt es im Feld eines Senders eine Zenneckwelle?. AEÜ 5 [1951], 15-24.

[2.46] Ott, H., Oberflächenwelle und kein Ende. AEÜ 5 [1951], 343-346.

[2.47] Kahan, T. and Eckart, G., Die Nichtexistenz der Oberflächenwelle in der Dipolstrahlung über ebener Erde. AEÜ 5 [1951], 25-32.

[2.48] Kahan, T. and Eckart, G., Zur Frage der Oberflächenwelle in der Dipolstrahlung über einer ebenen Erde. AEÜ 5 [1951], 347-348.

[2.49] Banõs, A., Dipole radiation in the presence of a conducting half-space. Pergamon Press, Oxford 1966.

[2.50] Riordan, J. and Sunde, E.D., Mutual impedance of grounded wires for horizontally stratified two-layer earth. Bell Syst. tech. J. 12 [1933], 162-177.

[2.51] Wolf, A., Electric field of an oscillating dipole on the surface of a two-layer earth. Geophysics 11 [1946], 518-534.

[2.52] Lewis, W.B., Discussion on electric field of an oscillating dipole on the surface of a two-layer earth. Geophysics 11 [1946], 535-537.

[2.53] Großkopf, J. and Vogt, K., Die Messung der elektrischen Leitfähigkeit bei geschichtetem Boden. Hochfrequenztechnik 58 [1941], 52-57.

[2.54] Großkopf, J., Über die Ausbreitung der Oberflächenwelle über geschichteten und gekrümmten Boden. Hochfrequenztechnik 58 [1941], 163-171.

[2.55] Großkopf, J., Das Strahlungsfeld eines vertikalen Dipolsenders über geschichtetem Boden. Hochfrequenztechnik 60 [1942], 136-141.

[2.56] Norton, K.A., The physical reality of space and surface waves in the radiation field of radio antennas. Proc. IRE 25 [1937], 1192-1202.

[2.57] Wait, J.R., Radiation from a vertical electric dipole over a stratified ground. Transact. IRE AP-1 [1953], 9-11.

[2.58] Wait, J.R. and Fraser, W.C.G., Radiation from a vertical dipole over a stratified ground (Part II). Transact. IRE AP-2 [1954], 144-146.

[2.59] Bhattacharyya, B.K., Electromagnetic induction in a two-layer earth. J. Geophys. Res. 60 [1955], 279-288.

[2.60] Slichter, L.B. and Knopoff, L., Field of an alternating magnetic dipole on the surface of a layered earth. Geophysics 24 [1959], 77-88.

[2.61] Wait, J.R., Influence of a sub-surface insulating layer on electromagnetic ground wave propagation. Transact. IEEE AP-14 [1966], 755-759.

[2.62] Wait, J.R., Asymptotic theory for dipole radiation in the presence of a lossy slab lying on a conducting half-space. Transact. IEEE AP-15 [1967], 645-648.

[2.63] Weaver, J.T., The quasi-static field of an electric dipole embedded in a two-layer conducting half-space. Can. J. Phys. 45 [1967], 1981-2002.

[2.64] Bannister, P.R., Summary of image theory expressions for the quasi-static fields of antennas at or above the earth's surface. Proc. IEEE 67 [1979], 1001-1008.

[2.65] Kuester, E.F. and Chang, D.C., Evaluation of Sommerfeld integrals associated with dipole sources above earth. Electromagn. Lab. Univ. of Colorado 43 [1979].

[2.66] Mahmoud, S.F. and Metwally, A.D., New image representation for dipoles near a dissipative earth 1. discrete images. Radio Sci. 16 [1981], 1271-1275.

[2.67] Thomson, D.J. and Weaver, J.T., Image approximation for quasistatic fields over a 2-layer conductor. Electronics Letters 6 [1970], 855-856.

[2.68] Ramaswamy, V., Dosso, H.W. and Weaver, J.T., Horizontal magnetic dipole embedded in a two-layer conducting medium. Can. J. Phys. 50 [1972], 607-616.

[2.69] Metwally, A.D. and Mahmoud, S.F., Mutual coupling between loops on layered earth using images. Transact. IEEE AP-32 [1984], 574-579.

[2.70] Mahmoud, S.F. and Mohsen, A.A., Assessment of image theory for field evaluation over a multilayer earth. Transact. IEEE AP-33 [1985], 1054-1057.

[2.71] Lindell, I.V. and Alanen, E., Exact image theory for the Sommerfeld half-space problem. Transact. IEEE AP-32 [1984]; part I: vertical magnetic dipole, 126-133; part II: vertical electric dipole, 841-847; part III: general formulation, 1027-1032.

[2.72] Alanen, E., Lindell, I.V. and Hujanen, A.T., Exact image method for field calculation in horizontally layered medium above a conducting ground plane. Proc. IEE Pt.H. 133 [1986], 297-304.

[2.73] Lindell, I.V., Alanen, E. and Hujanen, A.T., Exact image theory for the analysis of microstrip structures. Journal of Electromagnetic Waves and Applications 1 [1987], 95-108.

[2.74] Wait, J.R., Fields of a horizontal dipole over a stratified anisotropic half-space. Transact. IEEE AP-14 [1966], 790-792.

[2.75] Praus, O., Field of electric dipole above two-layer anisotropic medium. Studia Geophysica et Geodaetica 9 [1965], 359-380.

[2.76] Sinha, A.K. and Bhattacharya, P.K., Electric dipole over an anisotropic and inhomogenous earth. Geophysics 32 [1967], 652-667.

[2.77] Sinha, A.K., Vertical electric dipole over an inhomogenous and anisotropic earth. Pure and Appl. Geophys. 72 [1969], 123-147.

[2.78] Kong, J.A., Electromagnetic fields due to dipole antennas over stratified anisotropic media. Geophysics 37 [1972], 985-996.

[2.79] Stoyer, C.H., Electromagnetic fields of dipoles in stratified media. Transact. IEEE AP-25 [1977], 547-552.

[2.80] Cavalcante, G.P.S., Rogers, D.A. and Giarola, A.J., Analysis of electromagnetic wave propagation in multilayered media using dyadic Green's functions. Radio Sci. 17 [1982], 503-508.

[2.81] Kristensson, G., The electromagnetic field in a layered earth induced by an arbitrary stationary current distribution. Radio Sci. 18 [1983], 357-368.

[2.82] Wang, J.J.H., General method for the computation of radiation in stratified media. Proc. IEE Pt.H. 132 [1985], 58-62.

[2.83] Mohsen, A., Representation of the electromagnetic fields of a dipole above an inhomogeneous half space. Proc. IEE Pt.H. 132 [1985], 127-130.

[2.84] Sphicopoulos, T., Teodoridis, V. and Gardiol, F.E., Dyadic Green function for the elctromagnetic field in multilayered isotropic media: an operator approach. Proc. IEE Pt.H. 132 [1985], 329-334.

[2.85] Sami, M.A. and Samir, F.M., Electromagnetic fields of buried sources in stratified anisotropic media. Transact. IEEE AP-27 [1979], 671-678.

[2.86] Lee, J.K. and Kong, J.A., Dyadic Green's function for layered anisotropic medium. Electromagnetics 3 [1983], 111-130.

[2.87] Krowne, C.M., Green's function in the spectral domain for biaxial and uniaxial anisotropic planar dielectric structures. Transact. IEEE AP-32 [1984], 1273-1281.

[2.88] Tsalamengas, J.L. and Uzunoglu, N.K., Radiation from a dipole in the proximity of a general anisotropic grounded layer. Transact. IEEE AP-33 [1985], 165-172.

[2.89]   Felsen, L.B., Marcuvitz, N., Radiation and scattering of waves. Prentice-Hall,Inc. Englewood Cliffs, New Jersey 1973.

[2.90]   Papousek, W., Feldtypen der Dipolstrahlung in dreifach geschichteten Medien, Acta Physica Austr. 31 [1970], 109-123.

[2.91]   Papousek, W. and Schnizer, B., Surface impedance concepts of electromagnetic wave propagation in layered isotropic and anisotropic media. Radio Sci. 17 [1982], 1159-1167.

[2.92]   Gütl, A., Kemptner, E. and Papousek, W., Elektrischer Dipol über einer dielektrischen Schicht. Kleinheubacher Berichte 23 [1979], 135-143.

[2.93]   Tsang, L. and Kong, J.A., Interference patterns of a horizontal electric dipole over layered dielectric media. J. Geophys. Res. 78 [1973], 3287-3300.

[2.94]   Kong, J.A., Tsang, L. and Simmons, G., Geophysical subsurface probing with radio-frequency interferometry. Transact. IEEE AP-22 [1974], 616-620.

[2.95]   Tsang, L. and Kong, J.A., Electromagnetic fields due to a horizontal electric dipole antenna laid on the surface of a two-layer medium. Transact. IEEE AP-22 [1974], 709-711.

[2.96]   Cheng, D., Kong, J.A. and Tsang, L., Geophysical subsurface probing of a two-layered uniaxial medium with a horizontal magnetic dipole. Transact. IEEE AP-25 [1977], 766-769.

[2.97]   Chew, W.C. and Kong, J.A., Electromagnetic field of a dipole on a two-layer earth. Geophysics 46 [1981], 309-315.

[2.98]   Chew, W.C. and Kong, J.A., Asymptotic approximation of waves due to a dipole on a two-layer medium. Radio Sci. 17 [1982], 509-513.

[2.99]   Habashy, T.M., Kong, J.A. and Tsang, L., Quasi-static electromagnetic fields due to dipole antennas in bounded conducting media. Transact. IEEE GE-23 [1985], 325-333.

[2.100] Sommerfeld, A., Partielle Differentialgleichungen der Physik. Akademische Verlagsgesellschaft, Leipzig 1962.

[2.101] Keller, G.V. and Frischknecht, F.C., Electrical methods in geophysical prospecting. Pergamon Press, Oxford 1966.

[2.102] Wait, J.R., Electromagnetic waves in stratified media. Pergamon Press, Oxford 1970.

[2.103] King, R.W.P. and Smith, G.S., Antennas in matter. MIT Press, Cambridge 1981.

[2.104] Kong, J.A., Theory of electromagnetic waves. John Wiley & Sons, New York 1986.

[2.105] Gärtner, U., Lösung der Sommerfeldintegrale für eine ebene geschichtete Struktur mit Abschirmung. Diploma thesis, Institut für Hoch- und Höchstfrequenztechnik, RUB Bochum 1981.

[2.106] Siegel, M. and King, R.W.P., Electromagnetic fields in a dissipative half-space: A numerical approach. J. Appl. Phys. 41 [1970], 2415-2423.

[2.107] Bubenik, D.M., A practical method for the numerical evaluation of Sommerfeld integrals. Transact. IEEE AP-25 [1977], 904-906.

[2.108] Johnson, W.A. and Dudley, D.G., Real axis integration of Sommerfeld integrals: Source and observation points in air. Radio Sci. 18 [1983], 175-186.

[2.109] Lager, D.L. and Lytle, R.J., Numerical evaluation of Sommerfeld integrals. Lawrence Livermore Lab. Rep. UCRL-51821 [1975].

[2.110] Burke, G.J., Miller, E.K., Brittingham, J.N., Lager, D.L., Lytle, R.J. and Okada, J.T., Computer modeling of antennas near the ground. Electromagnetics 1 [1981], 29-49.

[2.111] Rahmat-Samii, Y., Mittra, R. and Parhami, P., Evaluation of Sommerfeld integrals for lossy half-space problems. Electromagnetics 1 [1981], 1-28.

[2.112] Michalski, K.A., On the efficient evaluation of integrals arising in the Sommerfeld halfspace problem. Proc. IEE Pt.H. 132 [1985], 312-318.

[2.113] Michalski, K.A., Smith, C.E. and Butler, C.M., Analysis of a horizontal two-element antenna array above a dielectric halfspace. Proc. IEE Pt.H. 132 [1985], 335-338.

[2.114] Katehi, P.B. and Alexopoulos, N.G., Real axis integration of Sommerfeld integrals with applications to printed circuit antennas. J. Math. Phys. 24 [1983], 527-533.

[2.115] Mahr, U., Analyse planarer Streifenleitungsantennen. Doctoral thesis, Universität Hagen - Gesamthochschule, [1983].

[2.116] Tsang, L., Brown, R., Kong, J.A. and Simmons, G., Numerical evaluation of electromagnetic fields due to dipole antennas in the presence of stratified media. J. Geophys. Res. 79 [1974], 2077-2080.

[2.117] Kuo, W.C. and Mei, K.K., Numerical approximations of the Sommerfeld integral for fast convergence. Radio Sci. 13 [1978], 407-415.

[2.118] Wait, J.R., Image theory of a quasistatic magnetic dipole over a dissipative half-space. Electronics Letters 5 [1969], 281-282.

References chapter 3

[3.1] Deschamps, G.A., Microstrip microwave antennas. 3rd USAF Symp. on Antennas [1953].

[3.2] Gutton, H. and Baissinot, G., Flat aerial for ultra high frequencies. French patent No. 703113 [1955].

[3.3] Lewin, L., Radiation from discontinuities in strip-line. Proc. IEE Pt. C 107 [1960], 163-170.

[3.4] Denlinger, E.J., Radiation from microstrip resonators. Transact. IEEE MTT-17 [1969], 235-236.

[3.5] Easter, B. and Roberts, R.J., Radiation from half-wavelength open-circuit microstrip resonators. Electronics Letters 6 [1970], 573-574.

[3.6] Koch, W., Die Strahlungsverluste von geraden, beidseitig leerlaufenden Microstripresonatoren. Nachrichtentechnik 22 [1972], 307-308.

[3.7] Easter, B. and Richings, J.G., Effects associated with radiation in coupled halfwave open-circuit microstrip resonators. Electronics Letters 8 [1972], 298-299.

[3.8] Sobol, H., Radiation conductance of open-circuit microstrip. Transact. IEEE MTT-19 [1971], 885-887.

[3.9] Byron, E.V., A new flush-mounted antenna element for phased array application. Proc. Phased-Array Antenna Symp. 1972, 187-192.

[3.10] Munson, R.E., Conformal microstrip antennas and microstrip phased arrays. Transact. IEEE AP-22 [1974], 74-78.

[3.11] Howell, J.Q., Microstrip antennas. Programme and Digest IEEE GAP-Int. Symp., Williamsburg Va. 1972, 177-180.

[3.12] James, J.R. and Wilson, G.J., Radiation characteristics of stripline antennas. European Microwave Conf., Montreux 1974, 484-488.

[3.13] Wiesbeck, W., Miniaturisierte Antenne in Mikrowellenstreifenleitungstechnik. NTZ 28 [1975], 156-159.

[3.14] Diener, G. und Entschladen, H., Messung der Strahlungseigenschaften von Mikrowellen-Streifenleitungs-Resonatoren. Frequenz 31 [1977], 55-59.

[3.15] Olthuis, R.W., Printed circuit antenna for radar altimeter. Int. Symp. IEEE AP-S, Massachusetts 1976, 554-557.

[3.16] James, J.R. and Wilson, G.J., Microstrip antennas and arrays. Pt. 1: Fundamental action and limitations, Pt. 2: New array-design technique. Microwave Optics and Acoustics 1 [1977], 165-181.

[3.17] Campbell, C.K., Traboulay, I., Suthers, M.S. and Kneve, H., Design of a stripline log-periodic dipole antenna, Transact. IEEE AP-25 [1977], 718-721.

[3.18] Rana, I.E., Alexopoulos, N.G. and Katehi, P.L., Theory of microstrip Yagi-Uda arrays. Radio Sci. 16 [1981], 1077-1079.

[3.19] Hall, P.S., New wideband microstrip antenna using log-periodic technique. Electronics Letters 16 [1980], 127-128.

[3.20] Hall, P.S., Bandwidth limitations of log-periodic microstrip patch antenna arrays. Electronics Letters 20 [1984], 437-438.

[3.21] Wood, C., Curved microstrip lines as compact wideband circularly polarised antennas. Microwave Optics and Acoustics 3 [1979], 5-13.

[3.22] Derneryd, A.G. and Karlsson, I., Broadband microstrip antenna element and array. Transact. IEEE AP-29 [1981], 140-141.

[3.23] Hall, P.S., Wood, C. and Garrett, C., Wide bandwidth microstrip antennas for circuit integration. Electronics Letters 15 [1979], 458-460.

[3.24] Wood, C., Improved bandwidth of microstrip antennas using parasitic elements. Proc. IEE Pt. H 127 [1980], 231-234.

[3.25] Kumar, G. and Gupta, K.C., Broad-band microstrip antennas using additional resonators gap-coupled to the radiating edges. Transact. IEEE AP-32 [1984], 1375-1379.

[3.26] Bhatnager, P.S., Daniel, J.-P., Mahdjoubi, K. and Terret, C., Experimental study on stacked triangular microstrip antennas. Electronics Letters 22 [1986], 864-865.

[3.27] Bhatnager, P.S., Daniel, J.-P., Mahdjoubi, K. and Terret, C., Hybrid edge, gap and directly coupled triangular microstrip antenna. Electronics Letters 22 [1986], 853-855.

[3.28] Aanandan, C.K. and Nair, K.G., Compact broadband microstrip antenna. Electronics Letters 22 [1986], 1064-1065.

[3.29] Dubost, G., Beauquet, G., Rocquencourt, J. and Bonnet, G., Patch antenna bandwidth increase by means of a director. Electronics Letters 22 [1986], 1345-1347.

[3.30] Menzel, W., A new travelling-wave antenna in microstrip. AEÜ 33 [1979], 137-140.

[3.31] Dong, W.-R. and Senggupta, D.L., A class of broad-band patch microstrip traveling wave antennas. Transact. IEEE AP-32 [1984], 98-100.

[3.32] Weinschel, H.D. and Carver, K.R., A medium-gain circularly polarized macrostrip UHF antenna for marine DCP communication to the GOES satellite system. Int. Symp. IEEE - APS, Massachusetts 1976, 391-394.

[3.33] McIlvenna, J. and Kernweis, N., Modified circular microstrip antenna elements. Electronics Letters 15 [1979], 207-208.

[3.34] Long, S.A., Shen, L.C., Schaubert, D.H. and Farrar, F.G., An experimental study of the circular-polarized elliptical printed-circuit antenna. Transact. IEEE AP-29 [1981], 95-98.

[3.35] Bahl, I.J. and Bhartia, P., Radiation characteristics of a triangular microstrip antenna. AEÜ 35 [1981], 214-219.

[3.36] Schaubert, D.H., Farrar, F.G., Sindoris, A. and Hayes, S.T., Microstrip antennas with frequency agility and polarization diversity. Transact. IEEE AP-29 [1981], 118-123.

[3.37] Shen, L.C., The elliptical microstrip antenna with circular polarization. Transact. IEEE AP-29 [1981], 90-94.

[3.38] Lo, Y.T. and Richards, W.F., Perturbation approach to design of circularly polarized microstrip antennas. Electronics Letters 17 [1981], 383-385.

[3.39] Sharma, P.C. and Gupta, K.C., Analysis and optimized design of single feed circularly polarized microstrip antennas. Transact. IEEE AP-31 [1983], 949-955.

[3.40] Huang, J., Circularly polarized conical patterns from circular microstrip antennas. Transact. IEEE AP-32 [1984], 991-994.

[3.41] Deshpande, M.D. and Das, N.K., Rectangular microstrip antenna for circular polarization. Transact. IEEE AP-34 [1986], 744-746.

[3.42] Carver, K.R. and Mink, J.W., Microstrip antenna technology. Transact. IEEE AP-29 [1981], 2-24.

[3.43] Mailloux, R.J., McIlvenna, J.F. and Kernweis, N.P., Microstrip array technology. Transact. IEEE AP-29 [1981], 25-37.

[3.44] James, J.R., Hall, P.S., Wood, C. and Henderson, A., Some recent developments in microstrip antenna design. Transact. IEEE AP-29 [1981], 124-128.

[3.45] Watkins, J., Radiation loss from open-circuited dielectric resonators. Transact. IEEE MTT-21 [1973], 636-639.

[3.46] Derneryd, A.G., Linearly polarized microstrip antennas. Transact. IEEE AP-24 [1976], 846-851.

[3.47] Knoppik, N., Der Gütefaktor von Mikrostrip-Resonatoren. AEÜ 30 [1976], 49-58.

[3.48] Wolff,I. und Knoppik, N., Mikrostrip-Scheibenresonatoren. AEÜ 28 [1974], 101-108.

[3.49] Derneryd, A.G., Analysis of the microstrip disk antenna element. Transact. IEEE AP-27 [1979], 660-664.

[3.50] Derneryd, A.G. and Lind, A.G., Extended analysis of rectangular microstrip resonator antennas. Transact. IEEE AP-27 [1979], 846-849.

[3.51] Malkomes, M. and Quitman, A., Microstrip circular-segment antenna. Electronics Letters 17 [1981], 198-200.

[3.52] Palanisamy, V. and Garg, R., Analysis of arbitrarily shaped microstrip patch antennas using segmentation technique and cavity model. Transact. IEEE AP-34 [1986], 1208-1213.

[3.53] Hammer, P., van Bouchaute, D., Verschraeven, D. and van de Capelle, A., A model for calculating the radiation field of microstrip antennas. Transact. IEEE AP-27 [1979], 267-270.

[3.54] Das, A., Das, S.K. and Mathur,S.P., Radiation characteristics of higher-order modes in microstrip ring antenna. Proc. IEE Pt. H 131 [1984], 102-106.

[3.55] James, J.R. and Henderson, A., High-frequency behaviour of microstrip open-circuit terminations. Microwaves Optics and Acoustics 3 [1979], 205-218.

[3.56] Kompa, G., Approximate calculation of radiation from open-ended wide microstrip lines. Electronics Letters 12 [1976], 222-224.

[3.57] Lier, E., Improved formulas for input impedance of coax-fed microstrip patch antennas. Proc. IEE Pt. H 129 [1982], 161-164.

[3.58] Lo, Y.T., Solomon, D. and Richards, W.F., Theory and experiment on microstrip antennas. Transact. IEEE AP-27 [1979], 137-145.

[3.59] Richards, W.F., Lo, Y.T. and Harrison, D.D., Improved theory for microstrip antennas. Electronics Letters 15 [1979], 42-44.

[3.60] Richards, W.F., Lo, Y.T. and Harrison, D.D., An improved theory for microstrip antennas and applications. Transact. IEEE AP-29 [1981], 38-46.

[3.61] Richards, W.F., Ou, J.-D. and Long, A., A theoretical and experimental investigation of annular, annular sector and circular sector microstrip antennas. Transact. IEEE AP-32 [1984], 864-867.

[3.62] Suzuki, Y. and Chiba, T., Computer analysis method for arbitrarily shaped microstrip antenna with multiterminals. Transact. IEEE AP-32 [1984], 585-590.

[3.63] Denlinger, E.J., Losses of microstrip lines. Transact. IEEE MTT-28 [1980], 513-522.

[3.64] Yano, S. and Ishimaru, A., A theoretical study of the input impedance of a circular microstrip disk antenna. Transact. IEEE AP-29 [1981], 77-83.

[3.65] Mink, J.W., Sensitivity of microstrip antennas to admittance boundary variations. Transact. IEEE AP-29 [1981], 142-144.

[3.66] De, A. and Das, B.N., Input impedance of probe-excited rectangular microstrip patch radiator. Proc. IEE Pt. H 131 [1984], 31-34.

[3.67] Itoh, T. and Mittra, R., A new method for calculating the capacitance of a circular disk for microwave integrated circuits. Transact. IEEE MTT-21 [1973], 431-432.

[3.68] Newman, E.H. and Tulyathan, P., Analysis of microstrip antennas using moment methods. Transact. IEEE AP-29 [1981], 47-53.

[3.69] Agrawal, P.K. and Bailey, M.C., An analysis technique for microstrip antennas. Transact. IEEE AP-25 [1977], 756-759.

[3.70] Itoh, T. and Menzel, W., A full-wave analysis method for open microstrip structures. Transact. IEEE AP-29 [1981], 63-67.

[3.71] Chew, W.C. and Kong, J.A., Analysis of a circular microstrip disk antenna with a thick dielectric substrate. Transact. IEEE AP-29 [1981], 68-76.

[3.72] Deshpande, M.D. and Bailey, M.C., Input impedance of microstrip antennas. Transact. IEEE AP-30 [1982], 645-650.

[3.73] Bailey, M.C. and Deshpande, M.D., Integral equation formulation of microstrip antennas. Transact. IEEE AP-30 [1982], 651-656.

[3.74] Pozar, D.M., Input impedance and mutual coupling of rectangular microstrip antennas. Transact. IEEE AP-30 [1982], 1191-1196.

[3.75] Araki, K. and Itoh, T., Hankel transform domain analysis of open circular microstrip radiating structures. Transact. IEEE AP-29 [1981], 84-89.

[3.76] Venkataraman, J. and Chang, D.C., Input impedance to a probe-fed rectangular microstrip patch antenna. Electromagnetics 3 [1983], 387-399.

[3.77] Pozar, D.M., Considerations for millimeter wave printed antennas. Transact. IEEE AP-31 [1983], 740-747.

[3.78] Bhattacharyya, A.K. and Garg, R., Spectral domain analysis of wall admittances for circular and annular microstrip patches and the effect of surface waves. Transact. IEEE AP-33 [1985], 1067-1073.

[3.79] Boukamp, J. and Jansen, R.H., Spectral domain investigation of surface wave excitation and radiation by microstrip lines and microstrip disk resonators. 13th European Microwave Conf. 1983.

[3.80] Pozar, D.M., Improved computational efficiency for the moment method solution of printed dipoles and patches. Electromagnetics 3 [1983], 299-309.

[3.81] Hansen, V. and Pätzold, M., Input impedance and mutual coupling of rectangular microstrip patch antennas with a dielectric cover. 16th European Microwave Conf., Dublin 1986, 643-648.

[3.82] Pozar, D.M. and Schaubert, D.H., Scan blindness in infinite phased arrays of printed dipoles. Transact. IEEE AP-32 [1984], 602-610.

[3.83] Pozar, D.M. and Schaubert, D.H., Analysis of an infinite array of rectangular microstrip patches with idealized probe feeds. Transact. IEEE AP-32 [1984], 1101-1107.

[3.84] Pozar, D.M., General relations for a phased array of printed antennas derived from infinite current sheets. Transact. IEEE AP-33 [1985], 498-504.

[3.85] Pozar, D.M., Finite phased arrays of rectangular microstrip patches. Transact. IEEE AP-34 [1986], 658-665.

[3.86] Eul, H.-J. and Hansen, V., Input impedance and radiation pattern of phased arrays of rectangular microstrip patch antennas with a dielectric cover. 17'th European Microwave Conf., Rom 1987.

[3.87] Uzunoglu, N.K., Alexopoulos, N.G. and Fikioris, J.G., Radiation properties of microstrip dipoles. Transact. IEEE AP-27 [1979], 853-858.

[3.88] Hansen, V., Calculation of Green's function for microstrip. 6th Antenna Symp., London 1980.

[3.89] Chew, W.C., Kong, J.A. and Shen, L.C., Radiation characteristics of a circular microstrip antenna. J. Appl. Phys. 51 [1980], 3907-3915.

[3.90] Mosig, J.R. and Gardiol, F.E., Current distribution in microstrip planar antennas. European Microwave Conf., Warschau 1980, 117-121.

[3.91] Hansen, V., Stromverteilung auf planaren Antennen über einer ebenen geschichteten Struktur. NTG-Fachtagung Antennen, Baden-Baden 1982, 58-62.

[3.92] Mosig, J.R. and Gardiol, F.E., Resonant frequency and input impedance of arbitrarily shaped microstrip antennas. Budapest 1982, 267-275.

[3.93] Mosig, J.R. and Gardiol, F.E., Untersuchung über beliebig geformte Mikrostreifenleitungsantennen unter Berücksichtigung von Oberflächenwelleneffekten. Mikrowellen Magazin 9 [1983], 423-425.

[3.94] Hansen, V., Berechnung der Anregung von Oberflächenwellen auf Streifenleitungen. Kleinheubacher Berichte 24 [1981], 45-49.

[3.95] De Assis Fonseca, S.B. and Giarola, A.J., Microstrip disk antennas, Pt. I: efficiency of space wave launching. Transact. IEEE AP-32 [1984], 561-567.

[3.96] De Assis Fonseca, S.B. and Giarola, A.J., Microstrip disk antennas, Pt. II: the problem of surface wave radiation by dielectric truncation. Transact. IEEE AP-32 [1984], 568-573.

[3.97] Katehi, P.B. and Alexopoulos, N.G., On the effect of substrate thickness and permittivity on printed circuit dipole properties. Transact. IEEE AP-31 [1983], 34-38.

[3.98] Alexopoulos, N.G., Katehi, P.B. and Rutledge, D.B., Substrate optimization for integrated circuit antennas. Transact. IEEE MTT-31 [1983], 550-557.

[3.99] Bhatthacharyya, A.K. and Garg, R., Effect of substrate on the efficiency of an arbitrarily shaped microstrip patch antenna. Transact. IEEE AP-34 [1986], 1181-1188.

[3.100] Alexopoulos, N.G. and Rana, I.E., Mutual impedance computation between printed dipoles. Transact. IEEE AP-29 [1981], 106-111.

[3.101] Hansen, V., Mutual coupling of dipoles in or above a stratified structure. Int. Symp. Digest IEEE AP-S 1983, 301-304.

[3.102] Newman, E.H., Richmond, J.H. and Kwan, B.W., Mutal impedance computation between microstrip antennas. Transact. IEEE MTT-31 [1983], 941-945.

[3.103] Wang, J. and Hansen, V., Streifenleitungsdipol mit dielektrischer Abdeckung. Kleinheubacher Berichte 26 [1983], 65-72.

[3.104] Alexopoulos, N.G. and Jackson, D.R., Radiation efficiency optimization for printed circuit antennas using magnetic superstrates. Electromagnetics 3 [1983], 255-269.

[3.105] Soares, A.J.M., de Assis Fonseca, S.B. and Giarola, A.J., The effect of a dielectric cover on the current distribution and input impedance of printed dipoles. Transact. IEEE AP-32 [1984], 1149-1153.

[3.106] Alexopoulos, N.G., Jackson, D.R. and Katehi, P.B., Criteria for nearly omnidirectional radiation patterns for printed antennas. Transact. IEEE AP-33 [1985], 195-205.

[3.107] Jackson, D.R. and Alexopoulos, N.G., Gain enhancement methods for printed circuit antennas. Transact. IEEE AP-33 [1985], 976-987.

[3.108] Hansen, V., Transient fields of a horizontal dipole on a plane stratified structure. Int. URSI Symp. on Electromag. Theory, Santiago de Compostella 1983, 203-206.

[3.109] Kediza, J.C. and Jecko, B., Frequency and time domain analysis of microstrip antennas. 15th European Microwave Conf., Paris 1985, 1045-1051.

[3.110] Wright, S.M. and Lo, Y.T., Efficient analysis for infinite microstrip dipole arrays. Electronics Letters 19 [1983], 1043-1045.

[3.111] Hansen, V. and Wang, J., Radiation coupling in a microstrip array with a dielectric cover. Proc. Int. Symp. on Antennas and EM Theory, Beijing 1985, 22-28.

[3.112] Newman, E.H. and Tehan, J.E., Analysis of a microstrip array and feed network. Transact. IEEE AP-33 [1985], 397-403.

[3.113] Pozar, D.M., Analysis of finite phased arrays of printed dipoles. Transact. IEEE AP-33 [1985], 1045-1053.

[3.114] Hansen, V., Finite array of printed dipoles with a dielectric cover. Accepted for Proc. IEE Pt. H.

[3.115] Hansen, V., Anregung von Leckwellen durch Mikrostreifenleitungsantennen. ITG Fachberichte 99 Antennen, Würzburg 1987, 251-255.

[3.116] Mosig, J.R. and Gardiol, F.E., Analytical and numerical techniques in the Green's function treatment of microstrip antennas and scatterers. Proc. IEE Pt. H 130 [1983], 175-182.

[3.117] Mosig, J.R. and Gardiol, F.E., An overview of integral equation techniques applied to microstrip antenna design. 15'th European Microwave Conf., Paris 1985, 1070-1077.

[3.118] Bahl, I.J. and Bhartia, P., Microstrip antennas. Artech House, Dedham MA, USA 1980.

[3.119] James, J.R., Hall, P.S. and Wood, C., Microstrip antenna theory and design. IEE Press, London 1981.

[3.120] Mosig, J.R. and Gardiol, F.E., A dynamical radiation model for microstrip structures. in: Advances in Electronics and Electron Physics, P.Hawkes ed., 59 [1982], 139-237.

References chapter 4

[4.1] Johnson, C.C. and Guy, A.W., Nonionizing electromagnetic wave effects in biological materials and systems. Proc. IEEE 60 [1972], 692-718.

[4.2] Short, J.G. and Turner, P.F., Physical hyperthermia and cancer therapy. Proc. IEEE 68 [1980], 133-142.

[4.3] Hahn, G.M., Hyperthermia for the engineer: A short biological primer. Transact. IEEE BME-31 [1984], 3-8.

[4.4] Schwan, H.P. and Foster, K.R., RF-field interactions with biological systems: Electrical properties and biophysical mechanisms. Proc. IEEE 68 [1980], 104-113.

[4.5] Lin, J.C., Microwave biophysics. Transact. Int. Microwave Power Inst. 8 [1978], 15-54.

[4.6] Schwan, H.P. and Piersol, G.M., The absorption of electromagnetic energy in body tissues. Rev. of Physical Medicine and Rehabilitation 33 [1954], 371-404.

[4.7] Schwan, H.P., Carstensen, E.L. and Li, K., Heating of fat-muscle layers by electromagnetic and ultrasonic diathermy. AIEE Pt.I 72 [1953], 483-488.

[4.8] Schwan, H.P., Radiation biology, medical applications, and radiation hazards. Microwave Power Engineering 2 [1968], 215-232.

[4.9] Lehmann, J.F., Guy, A.W., Johnston, V.C., Brunner, G.D. and Bell, J.W., Comparison of relative heating patterns produced in tissues by exposure to microwave energy at frequencies of 2450 and 900 megacycles. Arch. Phys. Med. Rehabil. 43 [1962], 69-76.

[4.10] Lehmann, J.F., Brunner, G.D., McMillan, J.A. and Guy, A.W., A comparative evaluation of temperature distributions produced by microwaves at 2456 and 900 megacycles in geometrically complex specimens. Arch. Phys. Med. Rehabil. 43 [1962], 502-507.

[4.11] Guy, A.W., Analyses of electromagnetic fields induced in biological tissues by thermographic studies on equivalent phantom models. Transact. IEEE MTT-19 [1971], 205-214.

[4.12] Nachman, M. and Turgeon, G., Heating pattern in a multi-layered material exposed to microwaves. Transact. IEEE MTT-32 [1984], 547-552.

[4.13] Guy, A.W., Electromagnetic fields and relative heating patterns due to a rectangular aperture source in direct contact with bilayered biological tissue. Transact. IEEE MTT-19 [1971], 214-223.

[4.14] Edenhofer, P., Field characteristics of a dual antenna sensor system probing biological tissues. Int. URSI Symp. on Electromagnetic Theory, Santiago de la Compostella 1983, 685-688.

[4.15] Guy, A.W., Lehmann, J.F., Stonebridge, J.B. and Sorensen, C.C., Development of a 915-MHz direct-contact applicator for therapeutic heating of tissues. Transact. IEEE MTT-26 [1978], 550-556.

[4.16] Lehmann, J.F., Guy, A.W., Stonebridge, J.B. and DeLateur, B.J., Evaluation of a therapeutic direct-contact 915-MHz microwave applicator for effective deep-tissue heating in humans. Transact. IEEE MTT-26 [1978], 556-563.

[4.17] Kantor, G., Witters, D.M. and Greiser, J.W., The peformance of a new direct contact applicator for microwave diathermy. Transact. IEEE MTT-26 [1978], 563-568.

[4.18] Stuchly, M.A., Stuchly, S.S. and Kantor, G., Diathermy applicators with circular aperture and corrugated flange. Transact. IEEE MTT-28 [1980], 267-271.

[4.19] Stuchly, S.S. and Stuchly, M.A., Multimode square waveguide applicators for medical applications of microwave power. Proc. 8th Europ. Microwave Conf., Paris 1978, 553-557.

[4.20] Lin, J.C., Kantor, G. and Ghods, A., A class of new microwave therapeutic applicators. Radio Sci. 17 [1982], 119S-123S.

[4.21] Paglione, R., Sterzer, F., Mendecki, J., Friedenthal, E. and Botstein, C., 27 MHz ridged waveguide applicators for localized hyperthermia treatment of deep-seated malignant tumors. Microwave J. 24 [1981], 71-80.

[4.22] Iskander, M.F. and Durney, C.H., An electromagnetic energy coupler for medical applications. Proc. IEEE 67 [1979], 1463-1465.

[4.23] Sterzer, F., Paglione, R., Nowogrodzki, M., Beck, E., Mendecki, J., Friedenthal, E. and Botstein, C., Microwave apparatus for the treatment of cancer. Microwave J. 23, No. 1 [1980], 39-44.

[4.24] Sterzer, F., Paglione, R., Mendecki, J., Friedenthal, E. and Botstein, C., RF therapy for malignancy. Spectrum IEEE 17 [1980], 32-37.

[4.25] Bahl, I.J., Stuchly, S.S. and Stuchly, M.A., New microstrip slot radiator for medical applications. Electronics Letters 16 [1980], 731-732.

[4.26] Johnson, R.H., James, J.R., Hand, J.W., Hopewell, J.W., Dunlop, P.R.C. and Dickinson,R.J., New low-profile applicators for local heating of tissues. Transact. IEEE BME-31 [1984], 28-37.

[4.27] Andersen, J.B., Baun, A., Harmark, K., Heinzl, L., Raskmark, P. and Overgaard, J., A hyperthermia system using a new type of inductive applicator. Transact. IEEE BME-31 [1984], 21-27.

[4.28] Henderson, A. and James, J.R., Near-field power transfer effects in small electromagnetic applicators for inducing hyperthermia. Proc. IEE Pt. H 132 [1985], 189-197.

[4.29] Johnson, R.H., New type of compact electromagnetic applicator for hyperthermia in the treatment of cancer. Electronics Letters 22 [1986], 591-593.

[4.30] Turner, P.F., Regional hyperthermia with an annular phased array. Transact. IEEE BME-31 [1984], 106-114.

[4.31] Raskmark, P. and Andersen, J.B., Focused electromagnetic heating of muscle tissue. Transact. IEEE MTT-32 [1984], 887-888.

[4.32] Turner, P.F., Mini-annular phased array for limb hyperthermia. Transact. IEEE MTT-34 [1986], 508-513.

[4.33] Sathiaseelan, V., Iskander, M.F., Howard, G.C.W. and Bleehen, N.M., Theoretical analysis and clinical demonstration of the effect of power pattern control using the annular phased-array hyperthermia system. Transact. IEEE MTT-34 [1986], 514-519.

[4.34] Anderson, A.P., Melek, M. and Brown, B.H., Feasibility of focused microwave array system for tumour irradiation. Electronics Letters 15 [1979], 564-565.

[4.35] Melek, M. and Anderson, A.P., Theoretical studies of localised tumour heating using focused microwave arrays. Proc. IEE Pt. F 127 [1980], 319-321.

[4.36] Cudd, P.A., Anderson, A.P., Hawley, M.S. and Conway, J., Phased-array design considerations for deep hyperthermia through layered tissue. Transact. IEEE MTT-34 [1986], 526-531.

[4.37] Arcangeli, G., Lombardini, P.P., Lovisolo, G.A., Marsiglia, G. and Piattelli, M., Focusing of 915 MHz electromagnetic power on deep human tissues: a mathematical model study. Transact. IEEE BME-31 [1984], 47-52.

[4.38] Livesay, D.E. and Chen, K.-M., Electromagnetic fields induced inside arbitrarily shaped biological bodies. Transact. IEEE MTT-22 [1974], 1273-1280.

[4.39] Jouvie, F., Bolomey, J.C. and Gaboriaud, G., Discussion of capabilities of microwave phased arrays for hyperthermia treatment of neck tumors. Transact. IEEE MTT-34 [1986], 495-501.

[4.40] Morita, N., Hamasaki, T. and Kumagai, N., An optimal excitation method in multi-applicator systems for forming a hot zone inside the human body. Transact. IEEE MTT-34 [1986], 532-538.

[4.41] Guy, A.W., Chou, C.-K. and Luk, K.H., 915-MHz phased-array system for treating tumors in cylindrical structures. Transact. IEEE MTT-34 [1986], 502-507.

[4.42] Gee, W., Lee, S.-W., Bong, N.K., Cain, C.A., Mittra, R. and Magin, R.L., Focused array hyperthermia applicator: Theory and experiment. Transact. IEEE BME-31 [1984], 38-46.

[4.43] Ling, H., Lee, S.-W. and Gee, W., Frequency optimization of focused microwave hyperthermia applicators. Proc. IEEE 72 [1984], 224-225.

[4.44] Hand, J.W., Cheetham, J.L. and Hind, A.J., Absorbed Power distributions from coherent microwave arrays for localized hyperthermia. Transact. IEEE MTT-34 [1986], 484-489.

[4.45] Kristensson, G., The electromagnetic field in a layered earth induced by an arbitrary stationary current distribution. Radio Sci. 18 [1983], 357-368.

[4.46] Galejs, J., Antennas in inhomogeneous media. Pergamon Press, Oxford 1969, 106.

[4.47] Clemmov, P.C., The plane wave spectrum representation of electromagnetic fields. Pergamon Press, Oxford 1966.

[4.48] Kaiser, M., Berechnung der Abstrahlung von Flächenstrahlern mit systemtheoretischen Methoden. Kleinheubacher Berichte 27 [1984], 137-145.

[4.49] Kaiser, M. and Hetsch, J., Derivation of an optimum focusing aperture illumination by a system-theoretic approach. Optica Acta 31 [1984], 225-232.

[4.50] Andersen, J.B., Electromagnetic heating. Proc. 4th Int. Symp. on Hyperthermic Oncology 1984, 113-128.

[4.51] Andersen, J.B., Theoretical limitations on radiation into muscle tissue. Int. J. Hyperthermia 1 [1985], 45-55.

[4.52] Loane, J., Ling, H., Wang, B.F. and Lee, S.W., Experimental investigation of a retro-focusing microwave hyperthermia applicator: conjugate-field matching scheme. Transact. IEEE-MTT-34 [1986], 490-494.

[4.53] Hand, J.W. and James, J.R., Physical techniques in clinical hyperthermia. Research Studies Press Ltd., Letchworth 1986.

[4.54] Wehner, R.S., Limitations of focused aperture antennas. USAF project RAND, Rand Corporation RM-262, Santa Monica 1949.

[4.55] Bickmore, R.W., On focusing electromagnetic radiators. Can. J. Phys. 35 [1957], 1292-1298.

[4.56] Sherman, J.W., Properties of focused apertures in the fresnel region. Transact. IRE AP-10 [1962], 399-408.

[4.57] Graham, W.J., Analysis and synthesis of axial field patterns of focused apertures. Transact. IEEE AP-31 [1983], 665-668.

[4.58] Hansen, R.C., Focal region characteristics of focused array antennas. Transact. IEEE AP-33 [1985], 1328-1337.

[4.59] Knoechel, R., Capabilities of multiapplicator systems for focussed hyperthermia. Transact. IEEE MTT-31 [1983], 70-73.

[4.60] Song, C.W., Lokshina, A., Rhee, J.G., Patten, M. and Levitt, S.H., Implication of blood flow in hyperthermic treatment of tumors. Transact. IEEE BME-31 [1984], 9-16.

[4.61] Bardati, F., Models of electromagnetic heating and radiometric microwave sensing. In: Physical techniques in clinical hyperthermia, Hand, J.W. and James, J.R. [4.53].

References chapter 5

[5.1] Habashy, T. and Mittra, R., On some inverse methods in electromagnetics. Journal of Electromagnetic Waves and Applications, 1 [1987], 25-58.

[5.2] Wait, J.R., Propagation of radio waves over a stratified ground. Geophysics [1953], 416-422.

[5.3] King, R.J., Crossed-dipole method of measuring wave tilt. Radio sci. 3 [1968], 345-350.

[5.4] King, R.J., On airborne wave tilt measurements. Radio sci. 12 [1977], 405-414.

[5.5] Maley, S.W., Radio wave methods for measuring the electrical parameters of the earth. In: Electromagnetic Probing in Geophysics, J.R. Wait, Ed., The Golem Press, Boulder, Colorado 1971, 77-95.

[5.6] King, R.J., Wave-tilt measurements. Transact. IEEE AP-24 [1976], 115-119.

[5.7] Hughes, W.J. and Wait, J.R., Effective wave tilt and surface impedance over a laterally inhomogeneous two-layer earth. Radio sci. 10 [1975], 1001-1008.

[5.8] Lytle, R.J., Lager, D.L. and Laine, E.F., Subsurface probing by high-frequency measurements of the wave tilt of electromagnetic surface waves. Transact. IEEE GE-14 [1976], 244-249.

[5.9] Riordan, J. and Sunde, E.D.,Mutual Impedance of Grounded Wires for Horizontally Stratified Two-Layer Earth. Bell Syst. Tech. Journ. 12 [1933], 162-177.

[5.10] Wait, J.R., The magnetic dipole over the horizontally stratified earth. Canadian Journal of Physics 29 [1951], 577-592.

[5.11] Wait, J.R., The fields of a line source of current over a stratified conductor. Appl. sci. Res., Sec. B 3 [1953], 279-292.

[5.12] El-Said, M.A.H., Geophysical prospecting of underground water in the desert by means of electromagnetic interference fringes. Proc. IRE 45 [1956], 24-30.

[5.13] Simmons, G.,Strangway, D.W., Bannister, L., Baker, R., Cubley, D., La-Torraca, G. and Watts, R., The surface electrical properties experiment. Lunar Geophysics, Reidel, Dordrecht, Netherlands [1972], 258.

[5.14] Simmons, G.,Strangway, D.W., Annan, P., Baker, R., Bannister, L., Brown, R., Cooper, W., Cubley, D., deBettencourt, J., England, A.W., Groener, J., Kong, J.A., LaTorraca, G., Meyer, J., Nanda, V., Redman, D., Rossiter, J., Tsang, L., Urner, J. and Watts, R., Surface electrical properties experiment. Apollo 17 preliminary science report, NASA, Spec. Publ. [1973], 15-1 – 15-13.

[5.15] Annan, A.P., Radio interferometry depth sounding. Part I - Theoretical discussion, Geophysics 38 [1973], 557-580.

[5.16] Tsang, L. and Kong, J.A., Interference pattern of a horizontal electric dipole over layered dielectric media. J. of Geophysical Research 78 [1973], 3287-3300.

[5.17] Tsang, G., Brown, R., Kong, J.A. and Simmons, G., Numerical evaluation of electromagnetic fields due to dipole antennas in the presence of stratified media. J. of Geophysical Research 79 [1974], 2077-2080.

[5.18] Kong, J.A., Tsang, L. and Simmons, G., Geophysical subsurface probing with radio-frequency interferometry. Transact. IEEE AP-22 [1974], 616-620.

[5.19] Fuller, J.A. and Wait, J.R., High-frequency electromagnetic coupling between small coplanar loops over an inhomogeneous Ground. Geophysics 37 [1972], 997-1004.

[5.20] Dey, A. and Ward, S.H., Inductive sounding of a layered earth with a horizontal magnetic dipole. Geophysics 35 [1970], 660-703.

[5.21] Rossiter, J.R., La Torraca, G.A., Annan, A.P.,Strangway, D.W. and Simmons, G., Radio interferometry depth sounding. Part II - Experimental results, Geophysics 38 [1973], 581-599.

[5.22] Wait, J.R., Electromagnetic Probing in Geophysics. The Golem Press, Boulder, Colorado 1971.

[5.23] Grant, F.S. and West, G.F., Interpretation theory in applied geophysics. McGraw-Hill, New York 1965.

[5.24] Tsang, L., Kong, J.A. and Shin, R.T., Theory of microwave remote sensing. John Wiley & Sons, New York 1985.

[5.25] Parkhomenko, E.I. (ed. by Keller, G.V.), Electrical properties of rocks. Plenum Press, New York 1967.

[5.26] Klein, A., Untersuchung der dielektrischen Eigenschaften feuchter Steinkohle im Hinblick auf die Anwendbarkeit des Mikrowellenverfahrens zur Wassergehaltsbestimmung. Doctoral thesis, Technische Hochschule Aachen [1978].

[5.27] Hallikainen, M.T., Ulaby, F.T., Dobson, M.C., El-Rayes, M.A. and Wu, L.-K., Microwave dielectric behaviour of wet soil - Part I: Empirical models and experimental observations. Transact. IEEE GE-23 [1985], 25-34.

[5.28] Dobson, M.C., Ulaby, F.T., Hallikainen, M.T. and El-Rayes, M.A., Microwave dielectric behavior of wet soil. Part II: Dielectric mixing models, Transact. IEEE GE-23 [1985], 35-46.

[5.29] Hoekstra, P. and Delaney, A., Dielectric properties of soils at UHF and microwave frequencies. J. of Geophysics Research 79 [1974], 1699-1708.

[5.30] Hipp, J.E., Soil electromagnetic parameters as a function of frequency, soil density and soil moisture. Proc. IEEE 62 [1974], 98-103.

[5.31] Wang, J.R. and Schmugge, T.J. An empirical model for the complex dielectric permittivity of soil as a function of water content. Trans. IEEE GE-18 [1980], 288-295.

[5.32] Wang, J.R., The dielectric properties of soil-water mixtures at microwave freqencies. Radio Sci. 15 [1980], 977-985.

[5.33] Schwarz, G., A theory of the low-frequency dielectric dispersion of colloidal particles in electrolyte solution. J. Phys. Chem. 66 [1962], 2636-2642.

[5.34] Wobschall ,D., A theory of the complex dielectric permittivity of soil containing water, the semidisperse model. IEEE GE-15 [1977], 49-58.

[5.35] de Loor, G.P., Dielectric properties of heterogeneous mixtures containing water. J. Microw. Power 3-2 [1968], 67-73.

[5.36] Vant, M.R., Ramseier, R.O., and Makios, V., The complex-dielectric constant of sea ice at frequencies in the range 0.1-40 GHz. J. Appl. Phys. 49 [1978], 1264-1280

[5.37] Cumming, W.A., The dielectric properties of ice and snow at 3.2 cm. J. Appl. Phys. 23 [1952], 768-773.

[5.38] von Hippel, A.R., Dielectric Materials and Applications. MIT Press, Cambridge, Mass. 1961.

[5.39] Evans, S., Dielectric properties of ice and snow - A review. J. Glaciol. 5 [1965], 773-792.

[5.40] Perry, J.W. and Straiton, A.W.,Dielectric constant of ice at 35.3 and 94.5 Ghz. J. Appl. Phys. 43 [1972], 731.

[5.41] Addison, J.R., Electrical relaxation in saline ice. J. Appl. Phys. 41 [1970], 54-63.

[5.42] Hoekstra, P., and Cappillino, P., Dielectric properties of sea and sodium chloride ice at UHF and microwave frequencies. J. Geophys. Res. 76 [1971], 4922.

[5.43] Vant, M.R., Gray, R.B., Ramseier, R.O., and Makios, V., Dielectric properties of fresh and sea ice at 10 and 35 GHz. J. Appl. Phys. 45 [1974], 4712-4717.

[5.44] Tobarias, J., Saguet, D. and Chilo, J., Determination of the water content of snow from the study of electromagnetic wave propagation in the snow cover. J. Glaciol. 20 [1980], 585-592.

[5.45] Linlor, W.I., Permittivity and attenuation of wet snow between 4 and 12 GHz. J. Appl. Phys. 51 [1980], 2811-2816.

[5.46] Glen, J.W. and Paren, P.G., The electrical properties of snow and ice. J. Glaciol. 15 [1975] , 15-38.

[5.47] von Hippel, A.R., Dielectrics and waves. MIT Press, 1966.

[5.48] Sadiku, M.N.O., Refractive index of snow at microwave frequencies. Appl. Opt. 24 [1985], 572-575.

[5.49] Ray, P.S., Broadband complex refractive indices of ice and water. Appl. Opt. 11 [1972], 1836-1844.

[5.50] Koizumi, S., Sato, K., Sato, T. and Shimba, M.,Dielectric characteristics of snow in microwave frequency. Electronics Letters 22 [1986], 823-825.

[5.51] Hallikainen, M.T., Ulaby, F.T. and Abdelrazik, M., Dielectric properties of snow in the 3 to 37 GHz range. Transact. IEEE AP-34 [1986], 1329-1339.

[5.52] Polder, D., Van Santen, J.H., The effective permeability of mixtures of solids. Physica 12 [1946], 257-271.

[5.53] Sinclair, G., Theory of models of electromagnetic systems. Proc. IRE 37 [1949], 1364-1370.

References chapter 6

[6.1] Manneback, C., Radiation from transmission lines. AIEE J. 42 [1923], 95-105.

[6.2] Schmitt, H.J., Transients in cylindrical antennae. Proc. IEE Monograph 377E [1960], 292-298.

[6.3] King, R.W.P., The theory of linear antennas. Harvard University Press, Cambridge 1956, 151-183.

[6.4] Brundell, P.O., Transient electromagnetic waves around a cylindrical transmitting antenna. Erricsson Tech. 16 [1959], 137-162.

[6.5] Einarsson, O., The step voltage current response of an infinite conducting cylinder. Trans. Roy. Inst. Technol. Stockholm 191 [1962], 2-14.

[6.6] Wu, T.T., Transient response of a dipole antenna. J. Math. Phys. 2 [1961], 892-894.

[6.7] Morgan, S.P., Transient response of a dipole antenna. J. Math. Phys. 3 [1962], 564-565.

[6.8] Latham, R.W. and Lee, K.S.H., Transient properties of an infinite cylindrical antenna. Radio Sci. 5 [1970], 715-723.

[6.9] Kasevich, R.S., Pulse response of linear dipole antenna. Transact. IEEE AP-31 [1983], 369-371.

[6.10] Wu, T.T., Theory of the dipole antenna and the two-wire transmission line. J. Math. Phys. 2 [1961], 550-574.

[6.11] King, R.W.P. and Schmitt, H.J., The transient response of linear antennas and loops. Transact. IRE AP-10 [1962], 222-228.

[6.12] Bolle, D.M. and Jacobs, I., The radiation pattern of long thin antennas for short-pulse excitation. Transact. IRE AP-10 [1962], 787-788.

[6.13] Schmitt, H.J., Harrison, C.W. and Williams, C.S., Calculated and experimental response of thin cylindrical antennas to pulse excitation. Transact. IEEE AP-14 [1966], 120-127.

[6.14] Harrison, C.W., The radian effective half-length of cylindrical antennas less than 1.3 wavelengths long. Transact. IEEE AP-11 [1963], 657-660.

[6.15] Harrison, C.W. and King, R.W.P., On the transient response of an infinite cylindrical antenna. Transact. IEEE AP-15 [1967], 301-302.

[6.16] Palciauskas, R.J. and Beam, R.E., Transient fields of thin cylindrical antennas. Transact. IEEE AP-18 [1970], 276-278.

[6.17] Hallen, E., Exact solution of the antenna equation. Transact. Roy. Inst. Technol. Stockholm 183 [1961], 2-22.

[6.18] Sato, M., Iguchi, M. and Sato, R., Transient response of coupled linear dipole antennas. Transact. IEEE AP-32 [1984], 133-140.

[6.19] Sato, M. and Sato, R., Analysis of transient responses between coupled dipole antennas by using a simple equivalent circuit. Transact. IEEE AP-33 [1985], 1015-1020.

[6.20] Landstorfer, F., Einschwingverhalten von Empfangsantennen. Doctaral-thesis, TH München 1967.

[6.21] Pozar, D.M., McIntosh, R.E. and Walker, S.G., The optimum feed voltage for a dipole antenna for pulse radiation. Transact. IEEE AP-31 [1983], 563-569.

[6.22] Pozar, D.M., Kang, Y.-W., Schaubert, D.H. and McIntosh, R.E., Opti-mization of the transient radiation from a dipole array. Transact. IEEE AP-33 [1985], 69-75.

[6.23] Abo-Zena, A.M. and Beam, R.E., Electromagnetic fields at points near pulse-excited linear antenna. Transact. IEEE AP-19 [1971], 129-131.

[6.24] Al-Badwaihy, K.A., Transient response of thick axially symmetric monopoles. Transact. IEEE AP-23 [1975], 428-431.

[6.25] Franceschetti, G. and Papas, C.H., Pulsed antennas. Transact. IEEE AP-22 [1974], 651-661.

[6.26] Sengupta, D.J. and Tai, C.-T., Radiation and reception of transients by linear antennas. In Transient electromagnetic fields. Ed. Felsen, L.B.,Springer Verlag, Berlin, New York 1976, 182-235.

[6.27] Wu, T.T. and King, R.W.P., The cylindrical antenna with nonreflecting resistive loading. Transact. IEEE AP-13 [1965], 369-373.

[6.28] Liu, Y.P. and Sengupta, D.L., Transient radiation from a linear antenna with nonreflecting resistive loading. Transact. IEEE AP-22 [1974], 212-220.

[6.29] Schuman, H., Time-domain scattering from a nonlinearly loaded wire. Transact. IEEE AP-22 [1974], 611-613.

[6.30] Bennett, C.L. and Weeks, W.L., Transient scattering from conducting cylinders. Transact. IEEE AP-18, [1970], 627-633.

[6.31] Sayre, E.P. and Harrington, R., Time domain radiation and scattering by thin wires. Appl. Sci. Res. 26 [1972], 413-444.

[6.32] Poggio, A.J., The space-time domain magnetic vector potential integral equations. Transact. IEEE AP-19 [1971], 702-704.

[6.33] Miller, E.K., Poggio, A.J. and Burke, G.J., An integro-differential equation technique for the time-domain analysis of thin-wire structures. J. Comput. Phys. 12 [1973], 24-48.

[6.34] Bennett, C.L. and Miller, E.K., Some computational aspects of transient electromagnetics. URSI spring meeting, Washington D.C. 1972, 109-110.

[6.35] Liu, T.K. and Mei, K.K., A time-domain integral-equation solution for linear antennas and scatterers. Radio Sci. 8 [1973], 797-804.

[6.36] Poggio, A.J. and Miller, E.K., Integral equation solutions of three-dimensional scattering problems. In Computer techniques for electromag-netics, Ed. Mittra, R., Pergamon Press, New York 1973, 159-264.

[6.37] Miller, E.K. and van Blaricum, M.L., The short-pulse response of a straight wire. Transact. IEEE AP-21 [1973], 396-398.

[6.38] Landt, J.A. and Miller, E.K., Transient response of the infinite cylindrical antenna and scatterer. Transact. IEEE AP-24 [1976], 246-250.

[6.39] Landt, J.A., Numerical analysis of transient fields near thin-wire antennas and scatterers. Radio Sci. 16 [1981], 1087-1091.

[6.40] Gomez, R., Morente, J.A. and Salinas, A., Time-domain analysis of an array of straight-wire coupled antennas. Electronics Letters 22 [1986], 316-318.

[6.41] Mittra, R., Integral equation methods for transient scattering. In Transient electromagnetic fields, Ed. Felsen, L.B., Springer Verlag, Berlin, New York 1976, 72-128.

[6.42] Miller, E.K. and Landt, J.A., Direct time-domain techniques for transient radiation and scattering from wires. Proc. IEEE 68 [1980], 1396-1423.

[6.43] Rao, S.M., Sarkar, T.K. and Dianat, S.A., The application of the conjugate gradient method to the solution of transient electromagnetic scattering from thin wires. Radio Sci. 19 [1984], 1319 - 1326.

[6.44] Rao, S.M., Sarkar, T.K. and Dianat, S.A., A novel technique to the solution of transient electromagnetic scattering from thin wires. Transact. IEEE AP-34 [1986], 630-634.

[6.45] Marin, L. and Latham, R.W., Representation of transient scattered fields in terms of free oscillations of bodies. Proc. IEEE 60 [1972], 640-641.

[6.46] Tesche, F.M., On the analysis of scattering and antenna problems using the singularity expansion technique. Transact. IEEE AP-21 [1973], 53-62.

[6.47] Marin, L., Natural-mode representation of transient scattered fields. Transact. IEEE AP-21 [1973], 809-818.

[6.48] Pine, Z.L. and Tesche, F.M., Calculation of the early time radiated electric field from a linear antenna with a finite source gap. Transact. IEEE AP-21 [1973], 740-743.

[6.49] Tesche, F.M., The far-field response of a step-excited linear antenna using SEM. Transact. IEEE AP-23 [1975], 834-838.

[6.50] Umashankar, K.R., Shumpert, T.H. and Wilton, D.R., Scattering by a thin wire parallel to a ground plane using the singularity expansion method. Transact. IEEE AP-23 [1975], 178-184.

[6.51] van Blaricum, M.L. and Mittra, R., A technique for extracting the poles and residues of a system directly from its transient response. Transact. IEEE AP-23 [1975], 777-781.

[6.52] Marin, L. and Liu, T.K., A simple way of solving transient thin-wire problems. Radio Sci. 11 [1976], 149-155.

[6.53] Liu, T.K. and Tesche, F.M., Analysis of antennas and scatterers with nonlinear loads. Transact. IEEE AP-24 [1976], 131-139.

[6.54] Miller, E.K., Brittingham, J.N. and Willows, J.L., Identification of E.M. spectrum by known pole sets. Electronics Letters 13 [1977], 774-775.

[6.55] Langenberg, K.J., Behandlung transienter Beugungs- und Abstrahlungssysteme mittels SEM (Singularity Expansion Method). Kleinheubacher

Berichte 21 [1978], 79.

[6.56] Langenberg, K.J., Transient fields of linear antenna arrays. Appl. Phys. 20 [1979], 101-118.

[6.57] Brittingham, J.N., Miller, E.K. and Willows, J.L., Pole extraction from real-frequency information. Proc. IEEE 68 [1980], 263-273.

[6.58] Baum, C.E., The singularity expansion method. In Transient electromagnetic fields, Ed. Felsen, L.B., Springer Verlag, Berlin, New York 1976, 130-179.

[6.59] Pearson, L.W. and Marin, L. (Guest Ed.), Special issue on the singularity expansion method. Electromagnetics 1 [1981], 349-512.

[6.60] van der Pol, B., On discontinuous electromagnetic waves and the occurrence of a surface wave. Transact. IRE AP-4 [1956], 288-293.

[6.61] Poritsky, H., Propagation of transient fields from dipoles near the ground. Br. J. Appl. Phys. 6 [1955], 421-426.

[6.62] Pekeris, C.L. and Alterman, Z., Radiation resulting from an impulsive current in a vertical antenna placed on a dielectric ground. J. Appl. Phys. 28 [1957], 1317-1323.

[6.63] Bremmer, H., The surface-wave concept in connection with propagation trajectories associated with the Sommerfeld Problem. Transact. IRE AP-7 [1959], S175-S182.

[6.64] Johler, J.R. and Walters, L.C., Propagation of a ground wave pulse around a finitely conducting spherical earth from a damped sinusoidal source current. Transact. IRE AP-7 [1959], 1-10.

[6.65] Wait, J.R., Transient fields of a vertical dipole over a homogeneous curved ground. Can. J. Phys. 34 [1956], 27-35.

[6.66] Wait, J.R., A note on the propagation of the transient ground wave. Can. J. Phys. 35 [1957], 1146-1151.

[6.67] Wait, J.R., Propagation of electromagnetic pulses in a homogeneous conducting earth. Appl. Sci. Res. (Sec. B) 8 [1959], 213-253.

[6.68] van der Pol, B. and Levelt, A.H.M., On the propagation of a discontinuous electromagnetic wave. Proc. Acad. Sci. Amsterdam Series A 63 [1960], 254-265.

[6.69] Frankena, H.J., Transient phenomena associated with Sommerfeld's horizontal dipole problem. Appl. Sci. Res. (Sec. B) 8 [1960], 357-368.

[6.70] De Hoop, A.T. and Frankena, H.J., Radiation of pulses generated by a vertical electric dipole above a plane, non- conducting, earth. Appl. Sci. Res. (Sec. B) 8 [1960], 369-377.

[6.71] Wait, J.R., A note on the propagation of electromagnetic pulses over the earth's surface. Can. J. Phys. 40 [1962], 1264-1269.

[6.72] Wait, J.R., On the electromagnetic pulse response of a dipole over a plane surface. Can. J. Phys. 52 [1973], 193-196.

[6.73] Fuller, J.A. and Wait, J.R., A pulsed dipole in the earth. In Transient electromagnetic fields. Ed. Felsen, L.B., Springer Verlag, Berlin , New

York 1976, 238-270.

[6.74] Lee, T., Transient electromagnetic waves applied to prospecting. Proc. IEEE 67 [1979], 1016-1021.

[6.75] Botros, A.Z. and Mahmoud, S.F., The transient fields of simple radiators from the point of view of remote sensing of the ground subsurface. Radio Sci. 13 [1978], 379-389.

[6.76] Haddad, H. and Chang, D.C., Transient electromagnetic field generated by a vertical electric dipole on the surface of a dissipative earth. Radio Sci. 16 [1981], 169-177.

[6.77] Ezzeddine, A., Kong, J.A. and Tsang, L., Transient fields of a vertical electric dipole over a two-layer nondispersive dielectric. J. Appl. Phys. 52 [1981], 1202-1208.

[6.78] Ezzeddine, A., Kong, J.A. and Tsang, L., Time response of a vertical electric dipole over a two-layer medium by the double deformation technique. J. Appl. Phys. 53 [1982], 813-822.

[6.79] Wait, J.R., The magnetic dipole over the horizontally stratified earth. Can. J. Phys. 29 [1951], 577-592.

[6.80] Wait, J.R., Current-carrying wire loops in a simple inhomogeneous region. J. Appl. Phys. 23 [1952], 497-498.

[6.81] Bhattacharyya, B.K., Electromagnetic fields of a transient magnetic dipole on the earth's surface. Geophysics 24 [1959], 89-108.

[6.82] Bhattacharyya, B.K., Electromagnetic fields of a vertical magnetic dipole placed above the earth's surface. Geophysics 28 [1963], 408-425.

[6.83] Wait, J.R. and Hill, D.A., Transient signals from a buried magnetic dipole. J. Appl. Phys. 42 [1971], 3866-3869.

[6.84] Wait, J.R. and Hill, D.A., Electromagnetic surface fields produced by a pulse-excited loop buried in the earth. J. Appl. Phys. 43 [1972], 3988-3991.

[6.85] Wait, J.R., On the theory of transient electromagnetic sounding over a stratified earth. Can. J. Phys. 50 [1972], 1055-1061.

[6.86] Mahmoud, S.F., Botros, A.Z. and Wait, J.R., Transient electromagnetic fields of a vertical magnetic dipole on a two-layer earth. Proc. IEEE 67 [1979], 1022-1029.

[6.87] Howard, A.Q., Transient response from a thin sheet in a conducting medium for loop excitation. Transact. IEEE GE-24 [1986], 198-203.

[6.88] Küster, E.F., The transient field reflected from a conducting half-space due to a pulsed electric or magnetic line current. Int. URSI Symp. on Electromagnetic Theory, Santiago de la Compostella 1983, 199-202.

[6.89] Küster, E.F. and Tijhuis, A.G., Two-dimensional transient scattering of an arbitrary electromagnetic field by a stratified dielectric and conducting region. Int. URSI Symp. on Electromagnetic Theory, Santiago de la Compostella 1983, 207-210.

[6.90] Kuester, E.F., The Transient electromagnetic field of a pulsed line source located above a dispersively reflecting surface. Transact. IEEE AP-32 [1984], 1154-1162.

[6.91] Chen, K.C., Time harmonic solutions for a long horizontal wire over the ground with grazing incidence. Transact. IEEE AP-33 [1985], 233-243.

[6.92] Degauque, P. and Thery, J.P., Electromagnetic subsurface radar using the Transient field radiated by a wire antenna. Transact. IEEE GE-24 [1986], 805-812.

[6.93] Rech, K.D., Becker, K.-D. and Langenberg, K.J., Effects of lossy and perfectly conducting ground on the time and frequency behaviour of a logperiodic antenna using the singularity expansion method. Radio Sci. 16 [1981], 1081-1086.

[6.94] Rahmat-Samii, Y., Parhami, P. and Mittra, R., Loaded horizontal antenna over an imperfect ground. Transact. IEEE AP-26 [1978], 789-796.

[6.95] Taylor, C.D. and Crow, T.T., Calculation of natural resonances for perpendicular crossed wires parallel to an imperfect ground. Electromagnetics 3 [1983], 41-64.

[6.96] Tesche, F.M., The effect of the thin-wire approximation and the source gap model on the high-frequency integral equation solution of radiating antennas. Transact. IEEE AP-20, [1972], 210-214.

[6.97] Mendel, J.M., Bremmer series decomposition of solutions to the lossless wave equation in layered media. Transact. IEEE GE-16 [1978], 103-112.

[6.98] Caspers, F., Bestimmung der Dicke und Dielektrizitätszahl ebener dielektrischer Schichten mit Hilfe von Freiraum - Impulsreflektometrie im Mikrowellenbereich. Doctoral-thesis, Ruhr-Universität Bochum 1982.

[6.99] Plehn, B., Transientes Verhalten von planaren Antennen. Diploma-thesis, Institut für Hoch- und Höchstfrequenztechnik, Ruhr-Universität Bochum 1982.

References appendix

[A1.1] Jones, D.S., Methods in electromagnetic wave propagation. Clarendon Press, Oxford 1979.

[A1.2] Harrington, R.F., Field computation by moment methods. The Macmillan Company, New York 1968.

[A1.3] Richmond, J.H., Radiation and scattering by thin-wire structures in the complex frequency domain. NASA Contractor Report CR-2396, Washington D.C. [1974], 1-39.

[A1.4] Rumsey, V.H., Reaction concept in electromagnetic theory. Phys. Rev. 94 [1954], 1483-1491.

[A1.5] Poggio, A.J. and Miller, E.K., Integral equation solutions of three-dimensional scattering problems. Computer techniques for electromagnetics, Ed. Mittra, R., Pergamon Press, Oxford 1973, 159-264.

[A1.6] Thiele, G.A., Wire antennas. Computer techniques for electromagnetics, Ed. Mittra, R., Pergamon Press, Oxford 1973, 7-95.

[A1.7] Miller, E.K. and Deadrick, F.J., Some computational aspects of thin-wire modeling. In Numerical and asymptotic techniques in electromagnetics, Ed. Mittra, R., Springer-Verlag, Berlin 1975.

[A1.8] Klein, C.A. and Mittra, R., The effect of different testing functions in the moment method solution of thin-wire antenna problems. Transact. IEEE AP-23 [1975], 258-261.

[A1.9] Thiele, G.A., Calculation of the current distribution on a thin linear antenna. Transact. IEEE AP-14 [1966], 648-649.

[A1.10] Harrington, R.F. and Mautz, J.R., Straight wires with arbitrary excitation and loading. Transact. IEEE AP-15 [1967], 502-515.

[A1.11] Unger, H.-G., Elektromagnetische Wellen I, Vieweg Verlag, Braunschweig 1967, 35.

Index